HEINEMANN MODULAR MATHEMATICS

for

EDEXCEL AS AND A-LEVEL

Revise for Core Mathematics 3

Alistair Macpherson Bronwen Moran Joe Petran
Keith Pledger Geoff Staley Dave Wilkins

Heinemann Educational Publishers
Halley Court, Jordan Hill, Oxford OX2 8EJ
Part of Harcourt Education

Heinemann is the registered trademark of
Harcourt Education Limited

First published 2005

09 08 07 06
10 9 8 7 6 5 4 3 2

British Library Cataloguing in Publication Data is available
from the British Library on request.

10-digit ISBN: 0 435 51125 4
13-digit ISBN: 978 0 43551 125 8

Designed by Bridge Creative Services
Typeset by Tech-Set Ltd

Illustrated by Tech-Set Ltd

Cover design by Bridge Creative Services

Printed and bound in China through Phoenix Offset

Acknowledgements
Every effort has been made to contact copyright holders of material reproduced in this book. Any omissions will be rectified in subsequent printings if notice is given to the publishers.

Tel: 01865 888058 email: info.he@heinemann.co.uk

About this book

This book is designed to help you get your best possible grade in your Core 3 examination. The authors are Chief and Principal examiners, and have a good understanding of Edexcel's requirements.

Revise for Core 3 covers the key topics that are tested in the Core 3 exam paper. You can use this book to help you revise at the end of your course, or you can use it throughout your course alongside the course textbook, *Heinemann Modular Mathematics for Edexcel AS and A-level Core 3*, which provides complete coverage of the syllabus.

Helping you prepare for your exam

To help you prepare, each topic offers you:

- **Key points to remember** – summarise the mathematical ideas you need to know and be able to use.
- **Worked examples and examination questions** – help you understand and remember important methods, and show you how to set out your answers clearly.
- **Revision exercises** – help you practise using these important methods to solve problems. Exam-level questions are included so you can be sure you are reaching the right standard, and answers are given at the back of the book so you can assess your progress.
- **Test Yourself questions** – help you see where you need extra revision and practice. If you do need extra help, they show you where to look in the *Heinemann Modular Mathematics for Edexcel AS and A-level Core 3* textbook and which example to refer to in this book.

Exam practice and advice on revising

Examination style paper – this paper at the end of the book provides a set of questions of examination standard. It gives you an opportunity to practise taking a complete exam before you meet the real thing. The answers are given at the back of the book.

How to revise – for advice on revising before the exam, read the How to revise section on the next page.

How to revise using this book

Making the best use of your revision time

The topics in this book have been arranged in a logical sequence so you can work your way through them from beginning to end. But **how** you work on them depends on how much time there is between now and your examination.

If you have plenty of time before the exam then you can **work through each topic in turn**, covering the key points and worked examples before doing the revision exercises and test yourself questions.

If you are short of time then you can **work through the Test Yourself sections** first, to help you see which topics you need to do further work on.

However much time you have to revise, make sure you break your revision into short blocks of about 40 minutes, separated by five- or ten-minute breaks. Nobody can study effectively for hours without a break.

Using the Test Yourself sections

Each Test Yourself section provides a set of key questions. Try each question:

- If you can do it and get the correct answer, then move on to the next topic. Come back to this topic later to consolidate your knowledge and understanding by working through the key points, worked examples and revision exercises.
- If you cannot do the question, or get an incorrect answer or part answer, then work through the key points, worked examples and revision exercises before trying the Test Yourself questions again. If you need more help, the cross-references beside each Test Yourself question show you where to find relevant information in the *Heinemann Modular Mathematics for Edexcel AS and A-level Core 3* textbook and which example in *Revise for C3* to refer to.

Reviewing the key points

Most of the key points are straightforward ideas that you can learn: try to understand each one. Imagine explaining each idea to a friend in your own words, and say it out loud as you do so. This is a better way of making the ideas stick than just reading them silently from the page.

As you work through the book, remember to go back over key points from earlier topics at least once a week. This will help you to remember them in the exam.

Algebraic fractions 1

Key points to remember

1 Algebraic fractions can be simplified by cancelling down. To do this the numerators and denominators must be fully factorised first.

2 If the numerator and denominator contain fractions then you can multiply both by the same number (the lowest common multiple) to create an equivalent fraction.

3 To multiply fractions, you simply multiply the numerators and multiply the denominators. If possible cancel down first.

4 To divide two fractions, multiply the first fraction by the reciprocal of the second fraction.

5 To add (or subtract) fractions each fraction must have the same denominator. This is done by finding the lowest common multiple of the denominators.

6 When the numerator has the same or higher degree than the denominator, you can divide the terms to produce a 'mixed' number fraction. This can be done either by using long division or by using the remainder theorem:

$$\mathrm{F}(x) \equiv \mathrm{Q}(x) \times \text{divisor} + \text{remainder}$$

where $\mathrm{Q}(x)$ is the quotient and is how many times the divisor divides into the function.

Example 1

Simplify: **(a)** $\dfrac{x^2 - 2x}{x^2 + x - 6}$

(b) $\dfrac{1 - \frac{1}{x}}{x - 1}$.

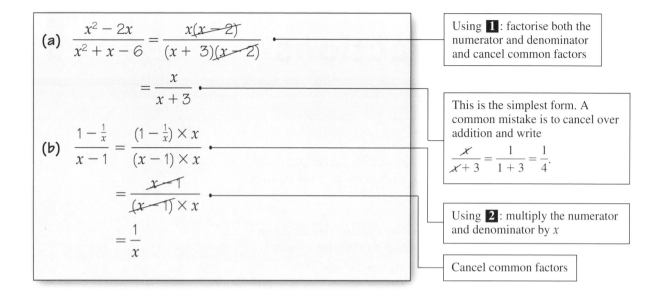

(a) $\dfrac{x^2 - 2x}{x^2 + x - 6} = \dfrac{x(x-2)}{(x+3)(x-2)}$

Using **1**: factorise both the numerator and denominator and cancel common factors

$= \dfrac{x}{x+3}$

This is the simplest form. A common mistake is to cancel over addition and write $\dfrac{x}{x+3} = \dfrac{1}{1+3} = \dfrac{1}{4}$.

(b) $\dfrac{1 - \frac{1}{x}}{x - 1} = \dfrac{(1 - \frac{1}{x}) \times x}{(x-1) \times x}$

Using **2**: multiply the numerator and denominator by x

$= \dfrac{x-1}{(x-1) \times x}$

Cancel common factors

$= \dfrac{1}{x}$

Example 2

Simplify: **(a)** $\dfrac{x}{y} \times \dfrac{y^2 - 3y}{x^2 + 5x}$ **(b)** $\dfrac{4}{x+2} \div \dfrac{2}{\frac{1}{2}x^2 + 2x + 2}$.

(a) $\dfrac{x}{y} \times \dfrac{y^2 - 3y}{x^2 + 5x} = \dfrac{x \times (y^2 - 3y)}{y \times (x^2 + 5x)}$

Using **3**

$= \dfrac{x \times y(y-3)}{y \times x(x+5)}$

Using **1**

$= \dfrac{y-3}{x+5}$

(b) $\dfrac{4}{(x+2)} \div \dfrac{2}{\frac{1}{2}x^2 + 2x + 2} = \dfrac{4}{(x+2)} \times \dfrac{\frac{1}{2}x^2 + 2x + 2}{2}$

Using **4**

$= \dfrac{4}{(x+2)} \times \dfrac{x^2 + 4x + 4}{4}$

Using **2**: multiply numerator and denominator of second fraction by 2

$= \dfrac{4}{(x+2)} \times \dfrac{(x+2)(x+2)}{4}$

Using **1**: factorise $x^2 + 4x + 4$ and cancel common factors

$= x + 2$

Example 3

Simplify the following: **(a)** $\dfrac{4}{x-1} + 3$ **(b)** $\dfrac{3}{(x+3)} - \dfrac{2}{(x+2)}$.

(a) $\dfrac{4}{x-1} + 3 = \dfrac{4}{(x-1)} + \dfrac{3}{1}$ — Write 3 as $\frac{3}{1}$

$= \dfrac{4}{(x-1)} + \dfrac{3(x-1)}{(x-1)}$ — Using **5**: the common denominator is $(x-1)$

$= \dfrac{4+3x-3}{(x-1)}$ — Simplify the numerator

$= \dfrac{3x+1}{x-1}$

(b) $\dfrac{3}{(x+3)} - \dfrac{2}{(x+2)}$

$\times (x+2)$ $\quad$ $\times (x+3)$

$= \dfrac{3(x+2)}{(x+3)(x+2)} - \dfrac{2(x+3)}{(x+3)(x+2)}$ — Using **5**: the common denominator is $(x+3)(x+2)$

$= \dfrac{3(x+2) - 2(x+3)}{(x+3)(x+2)}$

$= \dfrac{3x+6-2x-6}{(x+3)(x+2)}$ — Simplify the numerator

$= \dfrac{x}{(x+3)(x+2)}$

Worked exam style question 1

Show that $\dfrac{x^3 + 2x^2 - 5x + 7}{(x-1)} \equiv Ax^2 + Bx + C + \dfrac{D}{(x-1)}$, for constants A, B, C and D, which should be found.

NB. This question can be attempted using long division (see Heineman Book C3 page 7 Example 11) or by the remainder theorem.

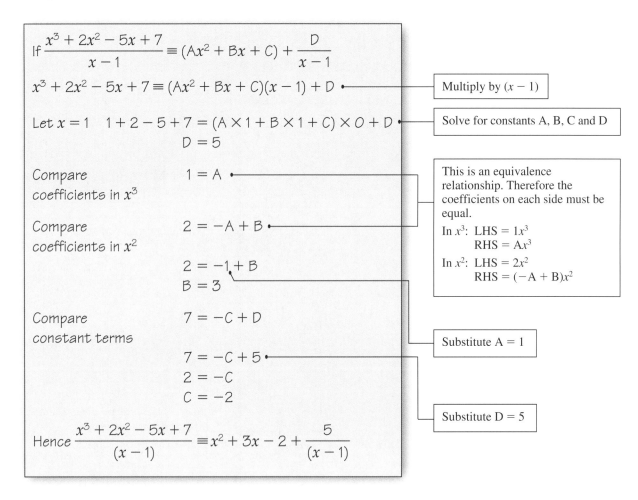

If $\frac{x^3 + 2x^2 - 5x + 7}{x - 1} \equiv (Ax^2 + Bx + C) + \frac{D}{x - 1}$

$x^3 + 2x^2 - 5x + 7 \equiv (Ax^2 + Bx + C)(x - 1) + D$ — Multiply by $(x - 1)$

Let $x = 1$ $\quad 1 + 2 - 5 + 7 = (A \times 1 + B \times 1 + C) \times 0 + D$ — Solve for constants A, B, C and D

$D = 5$

Compare coefficients in x^3 $\quad 1 = A$

Compare coefficients in x^2 $\quad 2 = -A + B$

This is an equivalence relationship. Therefore the coefficients on each side must be equal.

In x^3: LHS $= 1x^3$
RHS $= Ax^3$

In x^2: LHS $= 2x^2$
RHS $= (-A + B)x^2$

$2 = -1 + B$ — Substitute A = 1

$B = 3$

Compare constant terms $\quad 7 = -C + D$

$7 = -C + 5$ — Substitute D = 5

$2 = -C$

$C = -2$

Hence $\frac{x^3 + 2x^2 - 5x + 7}{(x - 1)} \equiv x^2 + 3x - 2 + \frac{5}{(x - 1)}$

Worked exam style question 2

$$f(x) = \frac{2x}{x + 2} + \frac{5x - 10}{x^2 - x - 6}, \qquad x > 3$$

(a) Show that $f(x) = 2 + \frac{1}{(x - 3)}, \qquad x > 3$

(b) Write down the range of f(x).

(a) $f(x) = \frac{2x}{(x + 2)} + \frac{5x - 10}{x^2 - x - 6}$

$= \frac{2x}{(x + 2)} + \frac{5x - 10}{(x + 2)(x - 3)}$ — Using **1**: factorise the denominator

$\times (x - 3)$ $\qquad \times 1$

$= \frac{2x(x - 3)}{(x + 2)(x - 3)} + \frac{5x - 10}{(x + 2)(x - 3)}$ — Using **5**: the common denominator is $(x + 2)(x - 3)$

$$= \frac{2x(x-3)+5x-10}{(x+2)(x-3)}$$ — Add the numerators

$$= \frac{2x^2-6x+5x-10}{(x+2)(x-3)}$$ — Simplify the numerator

$$= \frac{2x^2-x-10}{(x+2)(x-3)}$$

$$= \frac{(2x-5)\cancel{(x+2)}}{\cancel{(x+2)}(x-3)}$$ — Using **1**: factorise the numerator and cancel common factors

$$= \frac{2x-5}{(x-3)}$$

$$\begin{array}{r} 2 \\ x-3\overline{)2x-5} \\ \underline{2x-6} \\ 1 \end{array}$$

Divide $(x-3)$ into $(2x-5)$: it divides in 2 times

Multiply $(x-3)$ by 2 and subtract to find the remainder

Therefore $= 2 + \frac{1}{(x-3)}$

(b) $f(x) = 2 + \frac{1}{x-3}$

Hence Range of $f(x)$ is $f(x) > 2$ — When $x > 3$, $\frac{1}{x-3}$ is always positive

Revision exercise 1

1 Simplify the following fractions (if possible).

(a) $\frac{5y-5}{2y-2}$ **(b)** $\frac{a+6}{a+3}$ **(c)** $\frac{x^2+5x+6}{x^2+6x+8}$

2 Calculate the following:

(a) $\frac{x}{y} \times \frac{3}{x^2}$

(b) $\frac{3a^2}{5} \times \frac{10}{a^3}$

(c) $\frac{x+2}{x+4} \times \frac{x^2+5x+4}{x^2+3x+2}$.

3 Calculate the following:

(a) $\frac{a}{c} \div \frac{b}{c}$

(b) $\frac{4}{x^2} \div \frac{2}{x^3}$

(c) $\frac{x+3}{x-2} \div \frac{x^2+6x+9}{x^2-4}$.

4 Express the following improper fractions in 'mixed' number form.

(a) $\frac{x^3+2x^2-3x-5}{x-2}$ **(b)** $\frac{4x^3-2x^2+3x-6}{x^2+x-3}$

5 Show that $\frac{3x^3-5x^2-x+4}{x-2}$ can be put in the form

$Ax^2+Bx+C+\frac{D}{(x-2)}$.

Find the values of the constants A, B, C and D.

6 $f(x) = -x + \frac{x^2}{(x-2)} - \frac{20}{x^2+x-6}, x > 2$

(a) Show that $f(x) = \frac{2x+10}{x+3}, x > 2$.

(b) Hence show that the range of $f(x)$ is $2 < f(x) < 2.8$.

7 $f(x) = x + \frac{3}{x-1} - \frac{12}{x^2+2x-3}, x > 1$

(a) Show that $f(x) = \frac{x^2+3x+3}{x+3}$.

(b) Solve the equation $f'(x) = \frac{22}{25}$. *E*

Test yourself

What to review

If your answer is incorrect

1 Simplify if possible:

(a) $\dfrac{2a + 3}{\frac{1}{2}a + \frac{3}{4}}$ **(b)** $\dfrac{4y^2 - 9}{2y^2 + y - 3}$.

Review Heinemann Book C3 pages 1–2
Revise for C3 page 1 Example 1

---

2 Evaluate:

(a) $\dfrac{x^2 + 2x + 1}{y^3} \times \dfrac{y}{x^2 - 1}$ **(b)** $\dfrac{4b^2}{3} \div 2b$.

Review Heinemann Book C3 pages 3–4
Revise for C3 page 2 Example 2

---

3 Simplify:

(a) $\dfrac{4}{x - 1} + \dfrac{3x}{x^2 - 1}$ **(b)** $\dfrac{2}{(x + 2)} - \dfrac{3x}{(x + 2)^2}$.

Review Heinemann Book C3 pages 5–6
Revise for C3 page 3 Example 3

---

4 Express the following in mixed number form.

$$\frac{4x^3 - 5x^2 + 2x + 4}{x + 3}$$

Review Heinemann Book C3 pages 7–9
Revise for C3 page 3 Worked exam style question 1

---

5 $f(x) = x - 2 + \dfrac{4}{(x - 3)} - \dfrac{20}{x^2 - x - 6}, x > 3$

(a) Show that $f(x) = \dfrac{x^2}{x + 2}, x > 3$.

(b) Hence show that $f(x) > 1.8$.

Review Heinemann Book C3 page 6
Revise for C3 page 4 Worked exam style question 2

Test yourself answers

1 (a) 4 **(b)** $\dfrac{2y - 3}{y - 1}$ **2 (a)** $\dfrac{(x + 1)}{y^2(x - 1)}$ **(b)** $\dfrac{2b}{3}$

3 (a) $\dfrac{7x + 4}{(x - 1)(x + 1)}$ **(b)** $\dfrac{4 - x}{(x + 2)^2}$ **4** $4x^2 - 17x + 53 - \dfrac{155}{(x + 3)}$

Functions

2

Key points to remember

1 A function is a special mapping such that every element of the domain is mapped to exactly one element in the range.

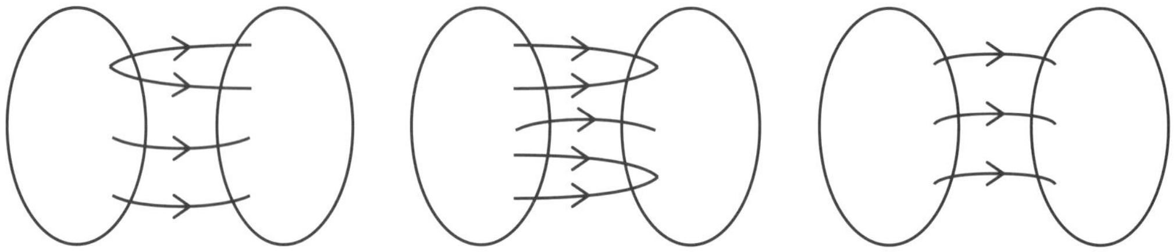

not a function | many-to-one function | one-to-one function

2 A function can be written in two different ways. For example, the function 'multiply by 3' can be expressed as

$f(x) = 3x \quad or \quad f: x \rightarrow 3x$

3 A one-to-one function is a special function where every element of the range is mapped from one element in the domain.

4 Many mappings can be made into functions by changing the domain. For example, the mapping 'positive square root' can be changed into the function $f(x) = \sqrt{x}$ by having a domain of $x \geqslant 0$.

5 If we combine two or more functions we can create a composite function. The function below is written $fg(x)$ as g acts on x first, then f acts on the result. For example,

$g(x) = 2x + 3,\ f(x) = x^2$
$fg(4) = f(2 \times 4 + 3) = f(11) = 11^2 = 121$

Similarly

$fg(x) = (2x + 3)^2$

6 The inverse of a function $f(x)$ is written $f^{-1}(x)$ and performs the opposite operation(s) to the function. To calculate the inverse function you change the subject of the formula. For example, the inverse function of $g(x) = 4x - 3$ is

$$g^{-1}(x) = \frac{x + 3}{4}$$

7 The range of the function is the domain of the inverse function and vice versa.

8 The graph of $f^{-1}(x)$ is a reflection of $f(x)$ in the line $y = x$.

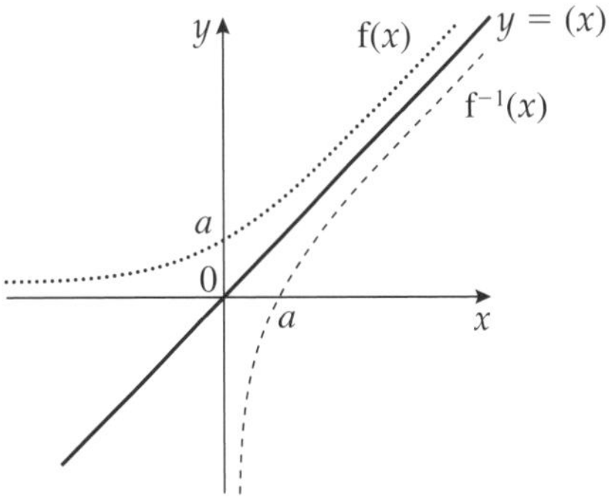

Example 1

$f(x) = 3x^2 - 1, x > 0$

Find: **(a)** f(4),
(b) the range of f(x),
(c) state why f(x) is a one-to-one function.

(a) $f(4) = 3 \times 4^2 - 1$
$= 48 - 1$
$= 47$

Substitute $x = 4$ into the formula

(b)

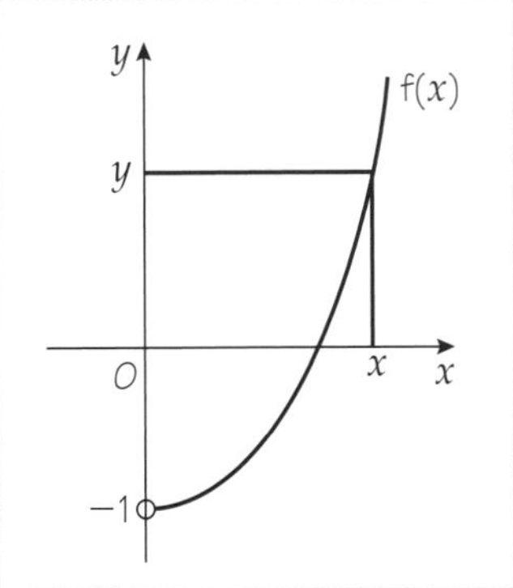

Sketch f(x) for $x > 0$, its domain. The range is the values that y take.

Range of f(x) is $f(x) > -1$

(c) f(x) is one-to-one because every y value has been mapped from a single value of x.

Using **3**: the definition of a one-to-one function

Example 2

$g(x) = x^2 - 2, x \in \mathbb{R}$

Find: **(a)** the values of b such that $g(b) = 14$,

(b) the values of c such that $g(c) = c$.

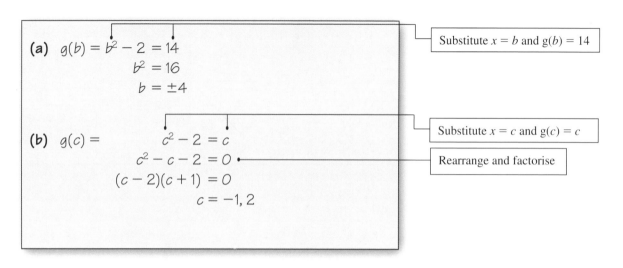

(a) $g(b) = b^2 - 2 = 14$

$b^2 = 16$

$b = \pm 4$

Substitute $x = b$ and $g(b) = 14$

(b) $g(c) = c^2 - 2 = c$

$c^2 - c - 2 = 0$

$(c - 2)(c + 1) = 0$

$c = -1, 2$

Substitute $x = c$ and $g(c) = c$

Rearrange and factorise

Example 3

The functions s and t are defined by $s : x \rightarrow 2x + 1, x \in \mathbb{R}$ and $t : x \rightarrow x^3, x \in \mathbb{R}$. Find:

(a) $st(2)$ **(b)** $ts(x)$ **(c)** $s^{-1}(x)$.

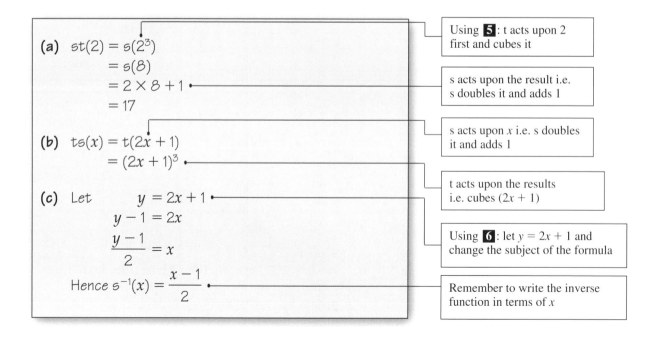

(a) $st(2) = s(2^3)$

$= s(8)$

$= 2 \times 8 + 1$

$= 17$

Using **5**: t acts upon 2 first and cubes it

s acts upon the result i.e. s doubles it and adds 1

(b) $ts(x) = t(2x + 1)$

$= (2x + 1)^3$

s acts upon x i.e. s doubles it and adds 1

t acts upon the results i.e. cubes $(2x + 1)$

(c) Let $y = 2x + 1$

$y - 1 = 2x$

$\frac{y - 1}{2} = x$

Hence $s^{-1}(x) = \frac{x - 1}{2}$

Using **6**: let $y = 2x + 1$ and change the subject of the formula

Remember to write the inverse function in terms of x

Example 4

$h(x) = x^2 + 5, x \geqslant 0$
Sketch the inverse function, $h^{-1}(x)$, stating its domain.

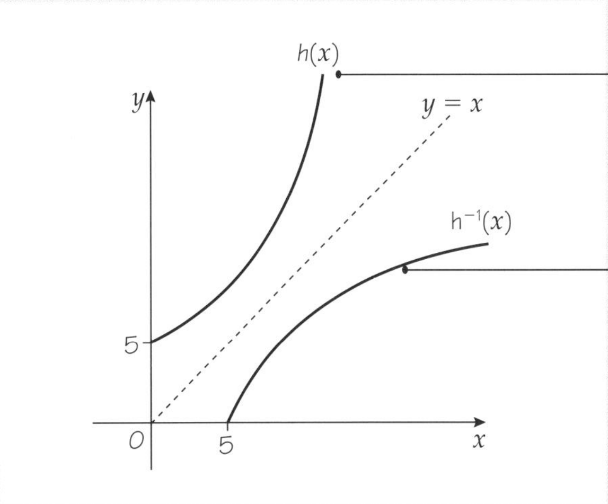

The domain of $h^{-1}(x)$ is $x \geqslant 5$

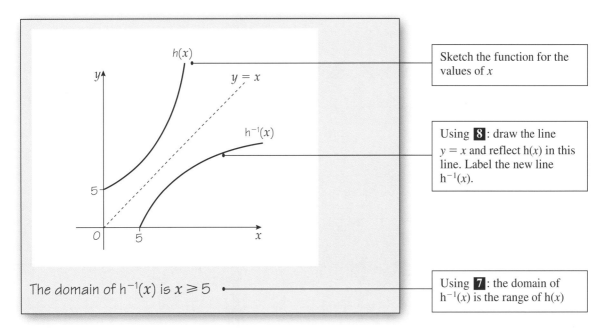

Sketch the function for the values of x

Using **8**: draw the line $y = x$ and reflect $h(x)$ in this line. Label the new line $h^{-1}(x)$.

Using **7**: the domain of $h^{-1}(x)$ is the range of $h(x)$

Worked exam style question 1

The function f is defined by

$$f : x \rightarrow x + 4 \qquad x > 0$$
$$f : x \rightarrow x^2 + 2 \qquad x \leqslant 0$$

(a) Sketch $f(x)$.

(b) State why $f(x)$ is **NOT** a one-to-one function.

(c) Find the values of x for which $f(x) = 20$.

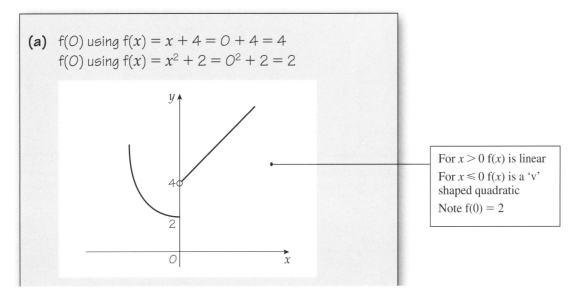

(a) $f(0)$ using $f(x) = x + 4 = 0 + 4 = 4$
$f(0)$ using $f(x) = x^2 + 2 = 0^2 + 2 = 2$

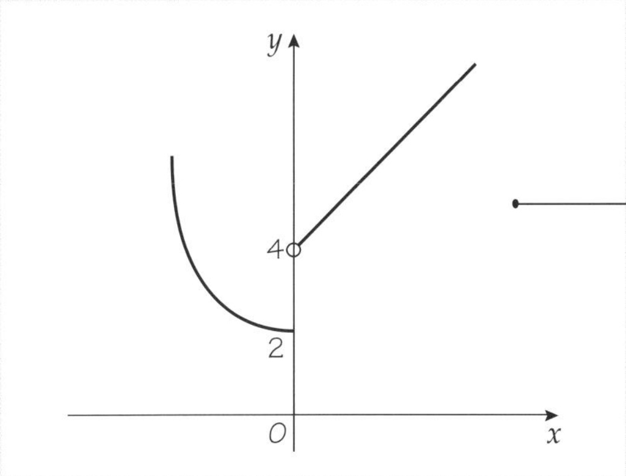

For $x > 0$ $f(x)$ is linear
For $x \leqslant 0$ $f(x)$ is a 'v' shaped quadratic
Note $f(0) = 2$

(b) f(x) is a MANY-TO-ONE function

It is **NOT** one-to-one because some y values are mapped from **two** different values of x (see below)

Using **3**: the definition of a one-to-one function

(c)

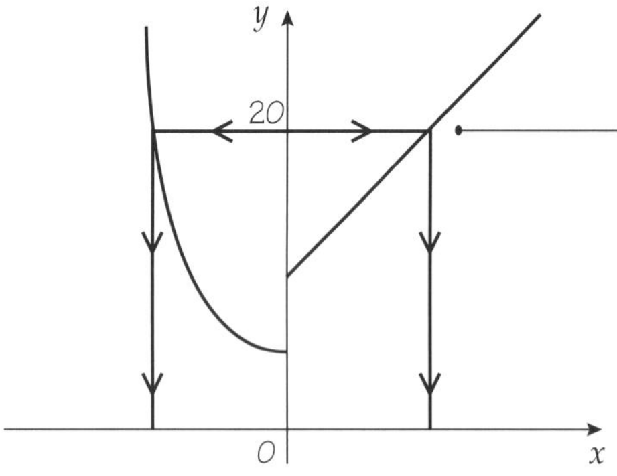

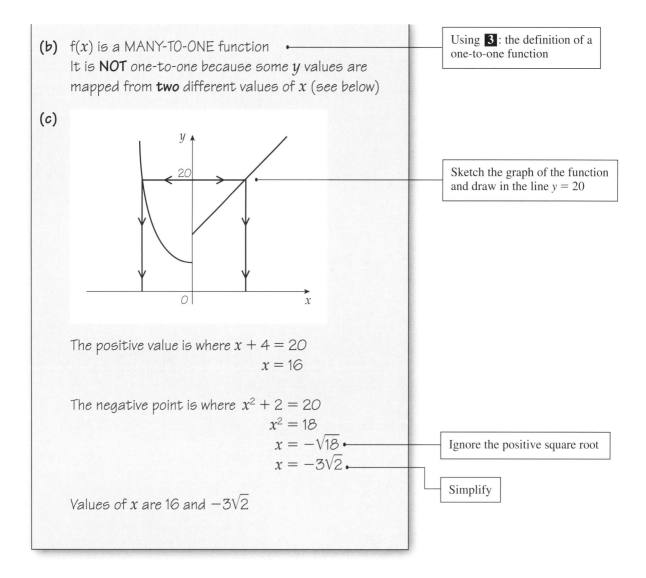

Sketch the graph of the function and draw in the line $y = 20$

The positive value is where $x + 4 = 20$

$$x = 16$$

The negative point is where $x^2 + 2 = 20$

$$x^2 = 18$$

$$x = -\sqrt{18}$$

Ignore the positive square root

$$x = -3\sqrt{2}$$

Simplify

Values of x are 16 and $-3\sqrt{2}$

Worked exam style question 2

The functions f and g are defined by

$$\mathrm{f}: x \rightarrow \frac{x}{x+3} \qquad \{x \in \mathbb{R}, x \neq -3\}$$

$$\mathrm{g}: x \rightarrow \frac{3}{x} \qquad \{x \in \mathbb{R}, x \geqslant 3\}$$

(a) Find an expression for $\mathrm{f}^{-1}(x)$.

(b) Find the range of $\mathrm{g}(x)$.

(c) Show that $\mathrm{fg}(x) = \dfrac{1}{1+x}$.

(a) Let $y = \dfrac{x}{x+3}$

$y(x+3) = x$ — Using **6**: rearrange to make x the subject of the formula

$yx + 3y = x$

$3y = x - yx$

$3y = x(1-y)$

$\dfrac{3y}{1-y} = x$

$f^{-1}(x) = \dfrac{3x}{1-x}$ — Define $f^{-1}(x)$ in terms of x

(b) $g(3) = \frac{3}{3} = 1$ — Calculate g(3)

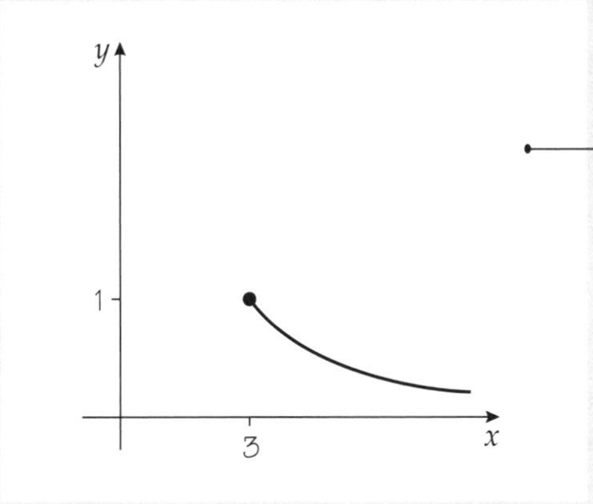

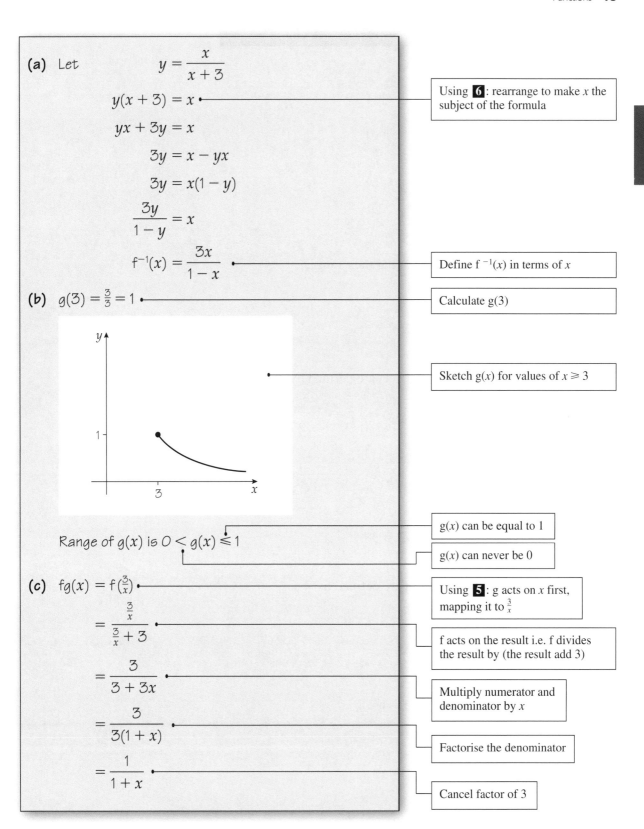

Sketch $g(x)$ for values of $x \geqslant 3$

Range of $g(x)$ is $0 < g(x) \leqslant 1$

$g(x)$ can be equal to 1

$g(x)$ can never be 0

(c) $fg(x) = f\left(\frac{3}{x}\right)$ — Using **5**: g acts on x first, mapping it to $\frac{3}{x}$

$= \dfrac{\frac{3}{x}}{\frac{3}{x}+3}$ — f acts on the result i.e. f divides the result by (the result add 3)

$= \dfrac{3}{3+3x}$ — Multiply numerator and denominator by x

$= \dfrac{3}{3(1+x)}$ — Factorise the denominator

$= \dfrac{1}{1+x}$ — Cancel factor of 3

Revision exercise 2

1 Find:

(a) $f(4)$ where $f(x) = 2^{x-1}$,

(b) $g(5)$ where $g(x) = \dfrac{x}{x-3}$.

2 Calculate a and b given that:

(a) $m(a) = 12$ where $m(x) = 4x + 5$,

(b) $n(b) = 1.6$ where $n(x) = \dfrac{2x}{x+2}$.

3 Sketch the following functions stating their range.

(a) $r(x) = x^2 - 2 \quad \{x \in \mathbb{R}, x > 0\}$

(b) $s(x) = x^3 \quad \{x \in \mathbb{R}, x \geqslant 0\}$

(c) $t(x) = x^2 - 10x + 30 \quad \{x \in \mathbb{R}\}$

4 The following mappings $f(x)$ and $g(x)$ are defined by

$$f(x) = \begin{cases} x^2 + 2 & x < 0 \\ 3x + 2 & x > 0 \end{cases} \qquad g(x) = \begin{cases} x^2 + 2 & x \leqslant 0 \\ 3x + 2 & x > 0 \end{cases}$$

Explain why $g(x)$ is a function and $f(x)$ is not.
Sketch $g(x)$ and find:

(a) $g(-2)$

(b) $g(3)$

(c) the value(s) of a such that $g(a) = 38$.

5 The following mappings m and n are defined by

$m : x \rightarrow 2x + 4$ and $n : x \rightarrow x^2 + 1$. Find:

(a) the function $mn(x)$,

(b) the function $nm(x)$,

(c) the function $m^2(x)$,

(d) the values of c such that $m(c) = n(c)$.

6 Find the inverses for each of:

(a) $f(x) = 2x + 5$

(b) $g(x) = 2x^3 - 5$

(c) $h(x) = \dfrac{1}{x+2} \quad \{x \in \mathbb{R}, x \neq -2\}$

7 The diagram below shows the function $p(x) = 2x^2 - 3, x \geqslant 0$.

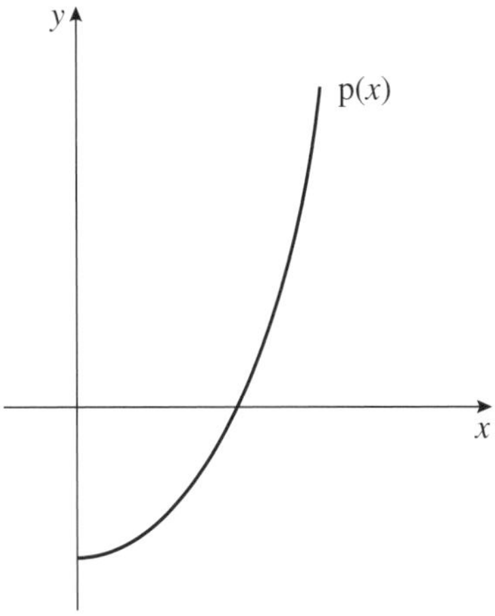

(a) State the range of $p(x)$.

(b) Calculate $p^{-1}(x)$ stating its domain.

(c) Sketch $p^{-1}(x)$ on the same set of axes as $p(x)$, stating the relationship between them.

(d) Hence or otherwise find the value of a such that $p(a) = p^{-1}(a)$.

8 The function g is defined by

$$g : x \rightarrow \frac{2x + 5}{x - 1} \quad \{x \in \mathbb{R}, x > 1\}$$

(a) Find $g^{-1}(x)$.

(b) Determine the range of $g^{-1}(x)$.

(c) Determine the domain of $g^{-1}(x)$.

9 The function f is defined by

$$f : x \rightarrow |x - 2| - 3 \quad \{x \in \mathbb{R}\}$$

(a) Solve the equation $f(x) = 1$.

The function g is defined by

$$g : x \rightarrow x^2 - 4x + 11, \quad x \geqslant 0$$

(b) Find the range of g.

(c) Find $gf(-1)$. *E*

Test yourself	What to review
	If your answer is incorrect
1 $f(x) = 2 \times 3^x - 1$ Find: **(a)** f(2), **(b)** the range of $f(x)$, **(c)** the value of b such that $f(b) = 161$.	*Review Heinemann Book C3 pages 12–15* *Revise for C3 pages 9–10* *Examples 1 and 2*
2 The function g is defined by $g : x \rightarrow x^2 + 5, x \geqslant 0$. **(a)** State the range of $g(x)$. **(b)** Determine the equation of $g^{-1}(x)$ stating its domain. **(c)** Sketch $g(x)$ and $g^{-1}(x)$ on the same set of axes stating the relationship between them.	*Review Heinemann Book C3 pages 15, 21–23* *Revise for C3 pages 10–11* *Examples 3 and 4*
3 If $f(x) = \dfrac{1}{x+2}$ and $g(x) = 3x - 2$, find in its simplest form: **(a)** $gf(x)$ **(b)** $gg(x)$ **(c)** $fg(x)$.	*Review Heinemann Book C3 pages 18–19* *Revise for C3 page 10* *Example 3*
4 The function n is defined by $n : x \rightarrow x^3 + 2 \quad x \in \mathbb{R}, x \geqslant 0$ $n : x \rightarrow 3 - 2x \quad x \in \mathbb{R}, x < 0$ **(a)** Sketch $n(x)$. **(b)** Determine $n(2)$ and $n(-2)$. **(c)** Calculate the values of x such that $n(x) = 29$. **(d)** State the range of $n(x)$.	*Review Heinemann Book C3 page 16* *Revise for C3 page 11* *Worked exam style question 1*
5 The function p is defined by $p : x \rightarrow \dfrac{4}{2x-1} \quad \{x \in \mathbb{R}, x \neq \frac{1}{2}\}$ **(a)** Find $p^{-1}(x)$ in its simplest form. **(b)** State the domain of $p^{-1}(x)$.	*Review Heinemann Book C3 page 22* *Revise for C3 page 12* *Worked exam style question 2*

Test yourself answers

1 (a) 17 **(b)** $f(x) > -1$ **(c)** 4

2 (a) $g(x) \geqslant 5$ **(b)** $g^{-1}(x) = \sqrt{x - 5}, x \geqslant 5$ **(c)**

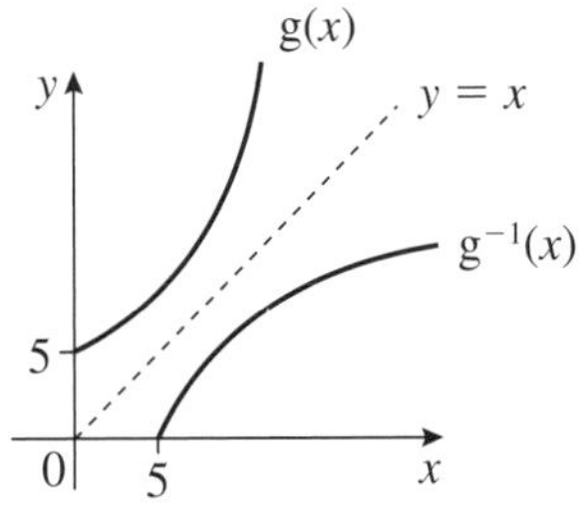

(reflection in $y = x$)

3 (a) $\frac{3}{(x + 2)} - 2$ **(b)** $9x - 8$ **(c)** $\frac{1}{3x}$

4 (a)

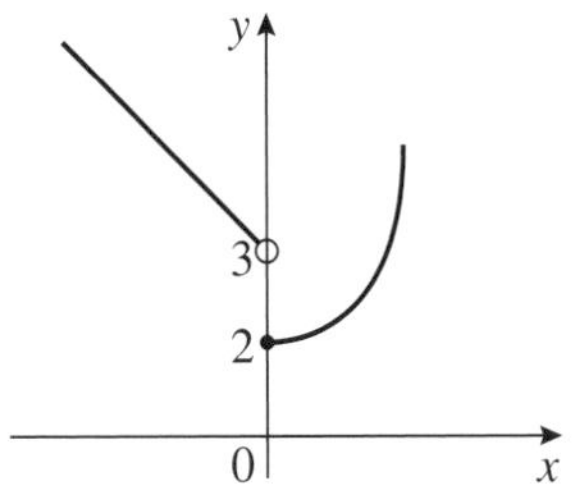

(b) 10, 7 **(c)** $-13, 3$ **(d)** $n(x) \geqslant 2$

3 (a) $p^{-1}(x) = \frac{x + 4}{2x}$ **(b)** $x \in \mathbb{R}, x \neq 0$

The Exponential and log functions

3

Key points to remember

1 Exponential functions are ones of the form $y = a^x$. They all pass through the point (0, 1).

The domain is all the real numbers.

The range is $f(x) > 0$.

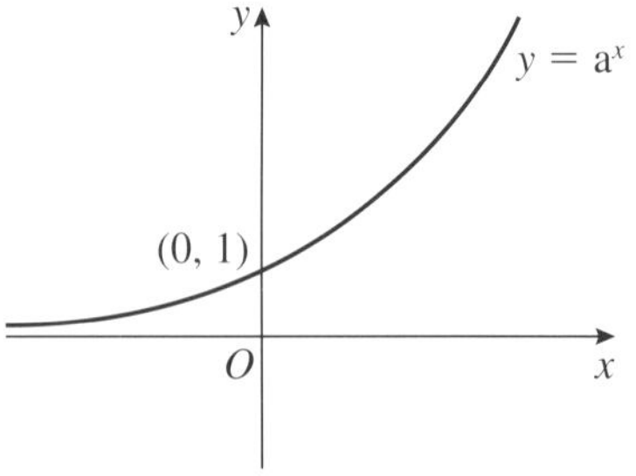

2 The exponential function $y = e^x$ (where $e \approx 2.718$) is a special function whose gradient is identical to the function. $[e^0 = 1]$

3 The inverse function to e^x is $\ln x$. $[\ln 1 = 0]$

4 The natural log function is a reflection of $y = e^x$ in the line $y = x$. It passes through the point (1, 0).

The domain is the positive numbers. The range is all the real numbers.

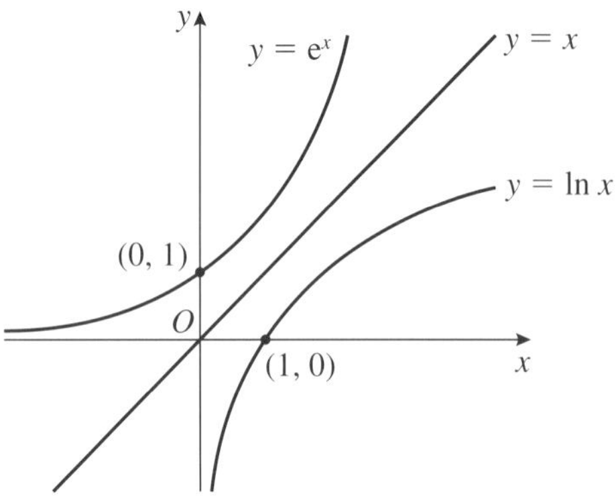

5 To solve an equation using $\ln x$ or e^x you must change the subject of the formula and use the fact that they are inverses of each other.

For example: (i) if $e^x = 5$, then $x = \ln 5$
(ii) if $\ln x = 2$, then $x = e^2$.

6 Growth and decay models are based around the exponential equations

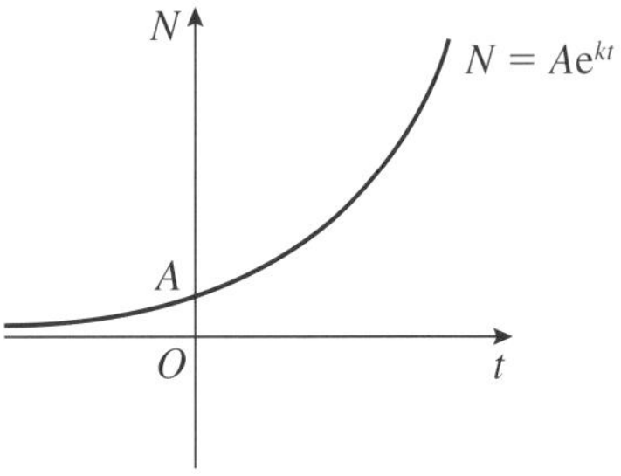

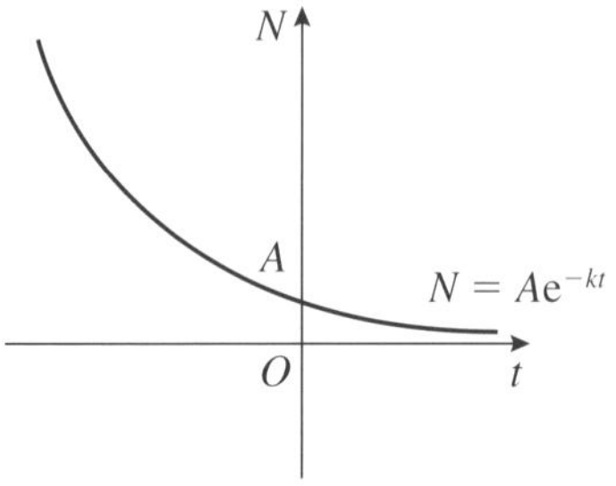

where A and k are positive numbers.

Example 1

The function f is defined over the set of real numbers by

$$f : x \rightarrow 3 + 2e^{\frac{1}{2}x}$$

(a) Find f(0) and f(6).

(b) Sketch the graph of the function f.

(c) State the range of f(x).

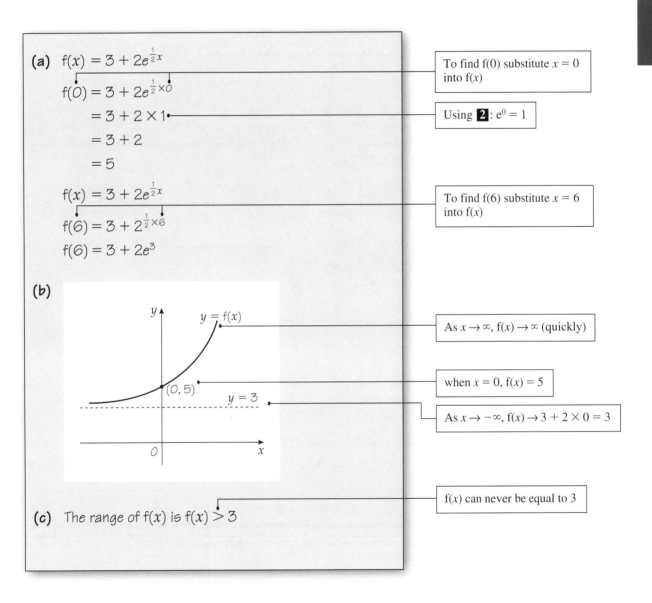

(a) $f(x) = 3 + 2e^{\frac{1}{2}x}$

$f(0) = 3 + 2e^{\frac{1}{2}\times 0}$

To find f(0) substitute $x = 0$ into f(x)

$= 3 + 2 \times 1$

Using **2**: $e^0 = 1$

$= 3 + 2$

$= 5$

$f(x) = 3 + 2e^{\frac{1}{2}x}$

$f(6) = 3 + 2^{\frac{1}{2}\times 6}$

To find f(6) substitute $x = 6$ into f(x)

$f(6) = 3 + 2e^3$

(b)

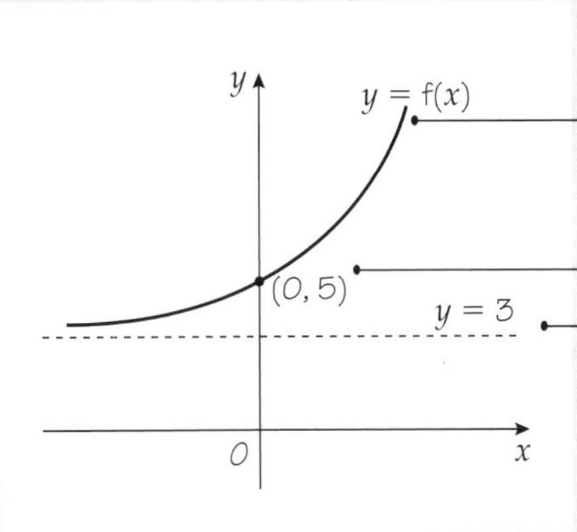

As $x \rightarrow \infty$, $f(x) \rightarrow \infty$ (quickly)

when $x = 0$, $f(x) = 5$

As $x \rightarrow -\infty$, $f(x) \rightarrow 3 + 2 \times 0 = 3$

(c) The range of f(x) is $f(x) > 3$

f(x) can never be equal to 3

Example 2

Give the exact solution to $\ln(4 - 2x) = 2$.

$$\ln(4 - 2x) = 2$$
$$4 - 2x = e^2$$
$$4 = e^2 + 2x$$
$$4 - e^2 = 2x$$
$$x = \frac{4 - e^2}{2}$$

Using **5** (ii): the inverse of ln is e

Make x the subject of the formula

As an exact answer is required there is no need to calculate this

Worked exam style question 1

The graph opposite shows the function f defined by

$f : x \rightarrow \ln(5x - 3) \quad \{x \in \mathbb{R}, x > \frac{3}{5}\}$

(a) State the co-ordinates of point A.

(b) Find an expression for $f^{-1}(x)$, stating its domain.

(c) Solve, giving your answer to 3 decimal places, $\ln(5x - 3) = 2$.

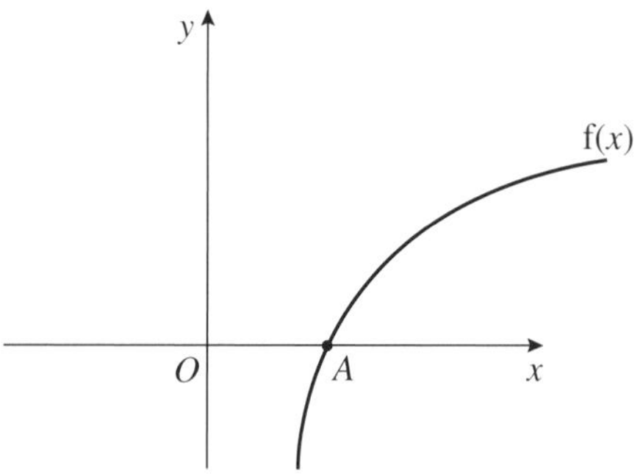

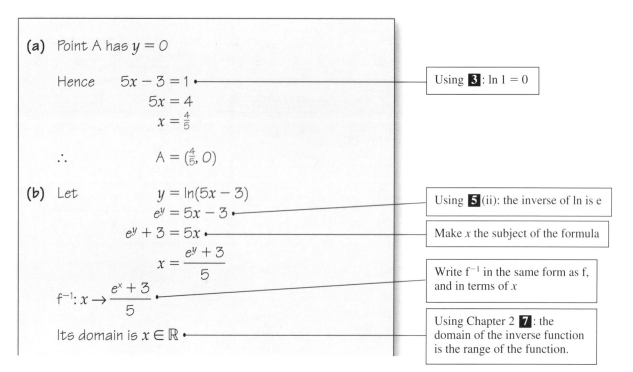

(a) Point A has $y = 0$

Hence $5x - 3 = 1$
$$5x = 4$$
$$x = \tfrac{4}{5}$$

$\therefore \quad A = (\tfrac{4}{5}, 0)$

Using **3**: $\ln 1 = 0$

(b) Let $y = \ln(5x - 3)$
$$e^y = 5x - 3$$
$$e^y + 3 = 5x$$
$$x = \frac{e^y + 3}{5}$$

$$f^{-1}: x \rightarrow \frac{e^x + 3}{5}$$

Its domain is $x \in \mathbb{R}$

Using **5** (ii): the inverse of ln is e

Make x the subject of the formula

Write f^{-1} in the same form as f, and in terms of x

Using Chapter 2 **7**: the domain of the inverse function is the range of the function.

(c) $\ln(5x - 3) = 2$

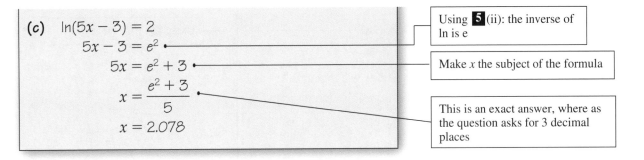

$5x - 3 = e^2$ — Using **5**(ii): the inverse of ln is e

$5x = e^2 + 3$ — Make x the subject of the formula

$x = \dfrac{e^2 + 3}{5}$ — This is an exact answer, where as the question asks for 3 decimal places

$x = 2.078$

Worked exam style question 2

The value of a car £C can be represented by the formula

$$C = 1000 + 17\,000\,e^{-\frac{t}{3}}$$

where t is the age in years from new.

(a) Calculate the price of the new car.

(b) Calculate the value of the car after 3 years.

(c) Calculate the age of the car when its value first falls below £3000.

(d) Find the car's value as $t \to \infty$. Explain why this is not a sensible answer.

(e) Sketch the graph showing C against t.

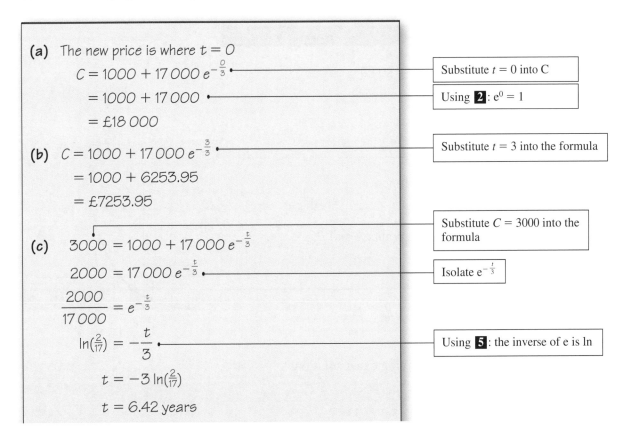

(a) The new price is where $t = 0$

$C = 1000 + 17\,000\,e^{-\frac{0}{3}}$ — Substitute $t = 0$ into C

$= 1000 + 17\,000$ — Using **2**: $e^0 = 1$

$= £18\,000$

(b) $C = 1000 + 17\,000\,e^{-\frac{3}{3}}$ — Substitute $t = 3$ into the formula

$= 1000 + 6253.95$

$= £7253.95$

(c) $3000 = 1000 + 17\,000\,e^{-\frac{t}{3}}$ — Substitute $C = 3000$ into the formula

$2000 = 17\,000\,e^{-\frac{t}{3}}$ — Isolate $e^{-\frac{t}{3}}$

$\dfrac{2000}{17\,000} = e^{-\frac{t}{3}}$

$\ln(\tfrac{2}{17}) = -\dfrac{t}{3}$ — Using **5**: the inverse of e is ln

$t = -3\ln(\tfrac{2}{17})$

$t = 6.42$ years

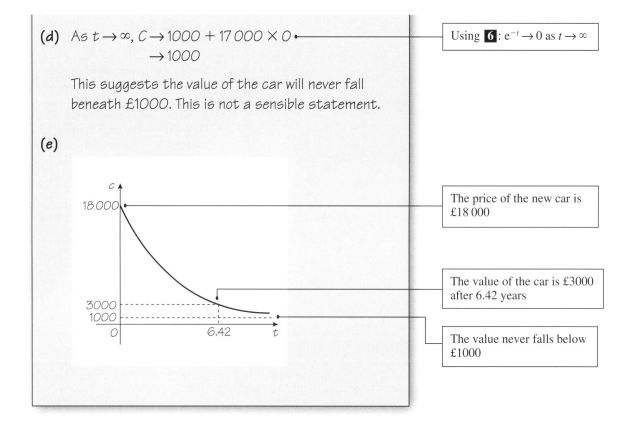

(d) As $t \to \infty$, $C \to 1000 + 17\,000 \times 0$
$\to 1000$

This suggests the value of the car will never fall beneath £1000. This is not a sensible statement.

Using **6**: $e^{-t} \to 0$ as $t \to \infty$

(e)

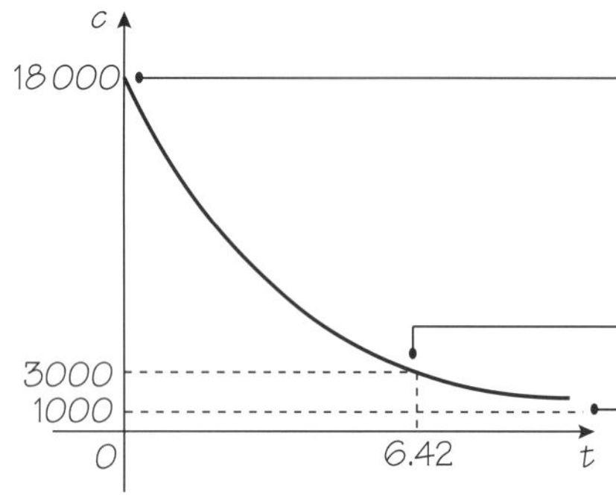

The price of the new car is £18 000

The value of the car is £3000 after 6.42 years

The value never falls below £1000

Revision exercise 3

1 Sketch the following 'exponential' graphs.

(a) $y = e^x + 3$

(b) $y = 2e^{-x}$

(c) $y = 5e^{\frac{1}{2}x} - 3$

2 The graph shows a sketch of the function

$$f(x) = 3e^{2x} - 5 \qquad \{x \in \mathbb{R}\}$$

(a) Find the exact coordinates of points A and B.

(b) State the range of $f(x)$.

(c) Find $f^{-1}(x)$, stating the domain.

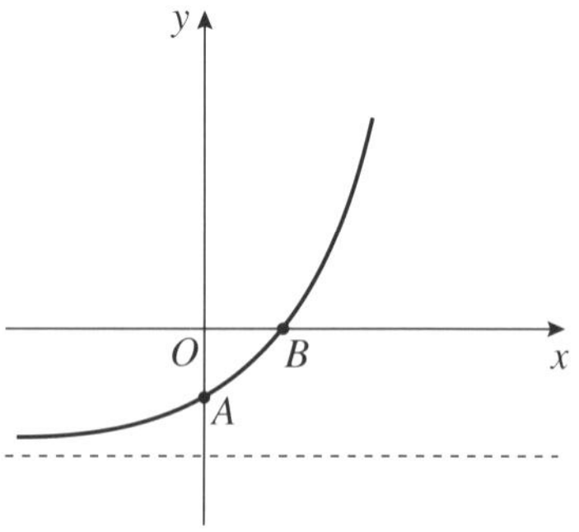

3 Solve the following equations, giving exact solutions.

(a) $2 \ln x + 1 = 4$ **(b)** $4e^{-2x} = 1$

(c) $\ln(2x + 3) = 8$ **(d)** $5e^x - 3 = 7$

4 The graph shows a sketch of the function

$f: x \to 3 + \ln(x + 5) \qquad \{x \in \mathbb{R}, x > -5\}$

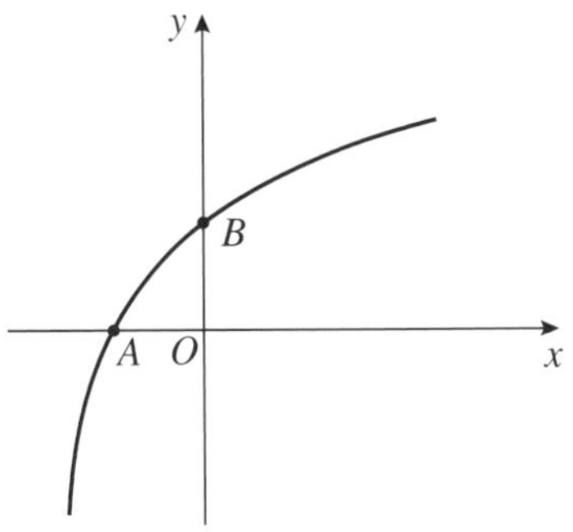

(a) Find the exact coordinates of points A and B.

(b) Express the inverse function $f^{-1}(x)$ in the form

$f^{-1}: x \to$

(c) Sketch, on the same set of axes, the graphs of $f(x)$ and $f^{-1}(x)$, labelling each carefully.

(d) Hence state the number of roots of equation $f(x) = f^{-1}(x)$.

5 The functions f and g are defined over the set of real numbers by

$f: x \to x - 5$
$g: x \to e^{-\frac{1}{2}x}$

(a) State the range of $g(x)$.
(b) Calculate $gf(5)$.
(c) Sketch the graphs of the inverse function f^{-1} and g^{-1}, marking on your sketches the coordinates of any points where your graphs cross the axes.

6 The function f is defined by

$f(x) = e^x - k, \quad x \in \mathbb{R}$ and k is a positive constant.

(a) State the range of $f(x)$.

(b) Find $f(\ln k)$, simplifying your answer.

(c) Solve the equation $f(2 \ln k) = 20$.

7 The number of seals on an island can be represented by the equation

$P = 200 - 120e^{-\frac{t}{10}}$

where P is the number of seals on the island and t is the time in years after January 1st 2005.

(a) Find the number of seals on the island on January 1st 2005.

(b) Calculate the number of extra seals on the island by January 1st 2010.

(c) Find the year when the population rises to over 180.

(d) Work out the number of seals that the island can sustain as predicted by the model.

8 The graph shows the function

$$h(x) = 80 - 10e^{5x}, \quad x \geqslant 0.$$

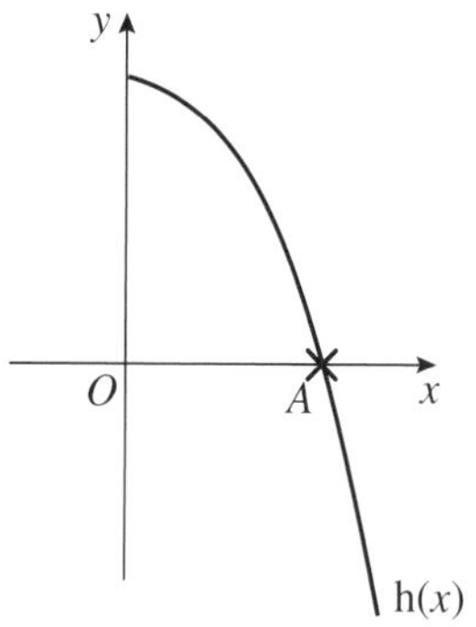

(a) State the range of $h(x)$.

(b) Calculate the coordinates of point A in terms of ln 2.

(c) Calculate $h^{-1}(x)$, stating its domain.

(d) Sketch, on the same set of axes, the graphs of $h(x)$ and $h^{-1}(x)$, stating the relationship between them.

9 As a substance cools its temperature, T °C, is related to the time (t minutes) for which it has been cooling. The relationship is given by the equation

$$T = 20 + 60e^{-0.1t}, \; t \geqslant 0.$$

(a) Find the value of T when the substance started to cool.

(b) Explain why the temperature of the substance is always above 20 °C.

(c) Sketch the graph of T against t.

(d) Find the value, to 2 significant figures, of t at the instant $T = 60$. **(E)**

Test yourself	What to review
	If your answer is incorrect
1 Sketch the graph of $f(x) = e^{2x} - 4$, indicating the points of intersection with both coordinate axes.	*Review Heinemann Book C3 pages 29–31* *Revise for C3 page 19 Example 1*
2 Solve the following equations giving answers to 3 d.p. **(a)** $3\ln(x + 2) = 5$ **(b)** $4e^{\frac{1}{2}x} - 1 = 10$	*Review Heinemann Book C3 page 35* *Revise for C3 page 20 Example 2*

3 The diagram shows the function where

$$g(x) = 2\ln(x + k)$$

where k is a constant, $k > 0$.

A and B are the points of intersection of $g(x)$ with the x- and y-axes respectively. Given the coordinates of B are $(0, \ln 9)$:

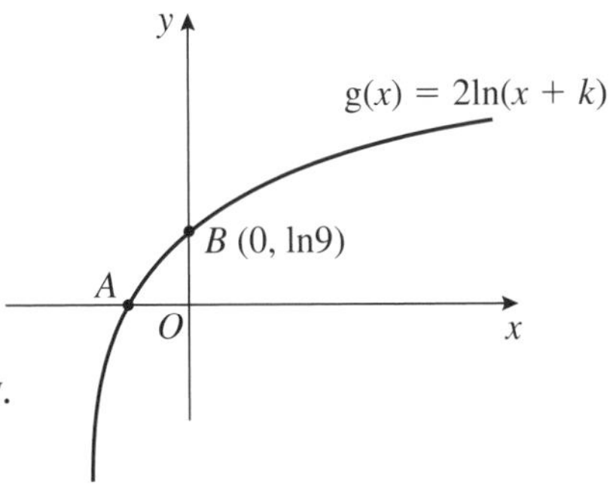

(a) find the exact value of k,

(b) determine the coordinates of A.

Review Heinemann Book C3 page 35
Revise for C3 page 20 Example 2 and Worked exam style question 1

---

4 The value of a used car can be represented by the equation

$$V = 15\,000e^{-\frac{1}{4}t}$$

where V = value of car in £'s and
t = the age of the car in years from new.

(a) State the price of the new car.

(b) Calculate the price of the car after 3 years.

(c) After how many years does the value fall below £5000.

Review Heinemann Book C3 pages 35–36
Revise for C3 page 21 Worked exam style question 2

---

5 The functions f and g are defined by

$$f(x) = e^x \quad \{x \in \mathbb{R}\} \qquad g(x) = \ln(x - k) \quad \{x \in \mathbb{R}, x > k\}$$

where k is a positive constant.

(a) State the range of f and the range of g.

(b) Calculate $fg(2k)$ in its simplest form.

(c) Given $gf(a) = 0$, determine 'a' in terms of k.

Review Heinemann Book C3 pages 30–35
Revise for C3 page 19 Example 1

Test yourself answers

1

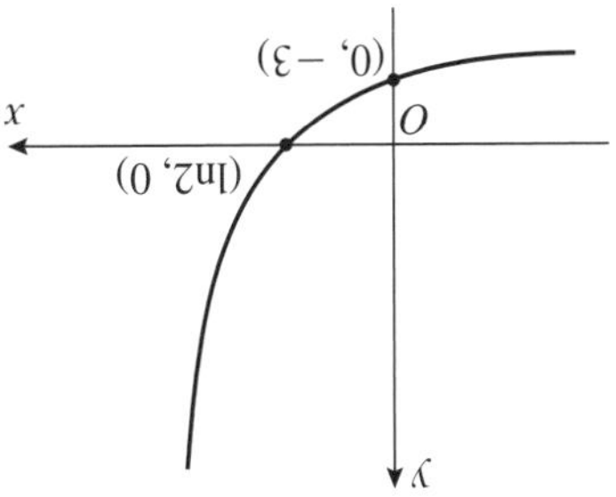

2 **(a)** $x = 3.294$ **(b)** 2.023 **3** **(a)** 3 **(b)** $(-2, 0)$

4 **(a)** £15 000 **(b)** £7085.50 **(c)** 4.4 years

5 **(a)** $f(x) > 0$, $g(x) \in \mathbb{R}$ **(b)** k **(c)** $a = \ln(1 + k)$

Numerical methods 4

Key points to remember

1 If you can find an interval in which f(x) changes sign, and f(x) is continuous in the interval, then the interval must contain a root of the equation f(x) = 0.

2 To solve an equation of the form f(x) = 0 by an iterative method, rearrange f(x) = 0 into a form $x = g(x)$ and use the iterative formula $x_{n+1} = g(x_n)$.

3 Different rearrangements of the equation f(x) = 0 give iterative formulae that **may** lead to different roots of the equation.

4 If you choose a value $x_0 = a$ for the starting value in an iterative formula, and $x_0 = a$ is close to a root of the equation f(x) = 0, then the sequence $x_0, x_1, x_2, x_3, x_4 \ldots$ does not necessarily converge to that root. In fact it might not converge to a root at all.

Example 1

Show that the equation $x^3 - \frac{3}{x} - 5 = 0$ has a root between $x = 1.8$ and $x = 1.9$.

Let $f(x) = x^3 - \frac{3}{x} - 5$

$$f(1.8) = (1.8)^3 - \frac{3}{1.8} - 5 = -0.835$$

$$f(1.9) = (1.9)^3 - \frac{3}{1.9} - 5 = 0.280$$

Show that the equation f(x) = 0 has a root in the interval $1.8 < x < 1.9$, by substituting $x = 1.8$ and $x = 1.9$ into the function.

$f(1.8) < 0$ and $f(1.9) > 0$, so there is a root between $x = 1.8$ and $x = 1.9$

Using **1**: there is a change of sign, so the equation f(x) = 0 has a root in the interval $1.8 < x < 1.9$.

Example 2

(a) On the same axes, sketch the graphs of $y = \sin x$ and $y = x^3 - 1$. Hence show that the equation $\sin x = x^3 - 1$ has only one root.

(b) Show that this root lies in the interval $1.2 < x < 1.3$.

(a)

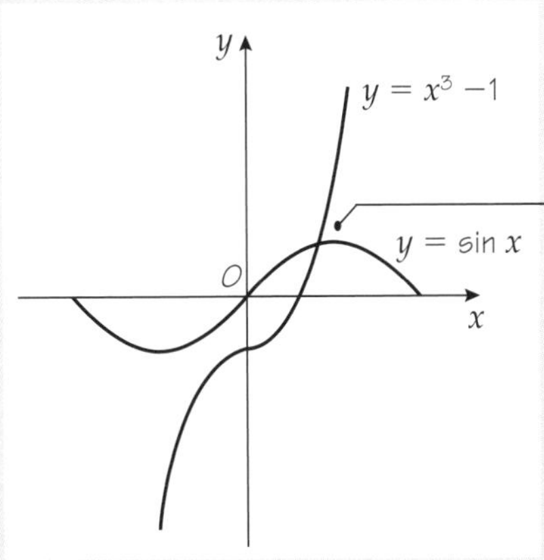

Draw $y = \sin x$ and $y = x^3 - 1$ on the same axes

The curves meet at only one point

As the curves meet only once, there is only one value of x that ssatisfies $\sin x = x^3 - 1$, so the equation $\sin x = x^3 - 1$ has only one root.

(b)

$$\sin x = x^3 - 1$$
$$\sin x + 1 = x^3$$
$$\sin x + 1 - x^3 = 0$$

Rearrange the equation into the form $f(x) = 0$

Let $f(x) = \sin x + 1 - x^3$

$$f(1.2) = \sin(1.2) + 1 - (1.2)^3 = 0.204$$

$$f(1.3) = \sin(1.3) + 1 - (1.3)^3 = -0.233$$

Show that the equation $f(x) = 0$ has a root in the interval $1.2 < x < 1.3$, by substituting $x = 1.2$ and $x = 1.3$ into the function.

$f(1.2) > 0$ and $f(1.3) < 0$, so the root lies in the interval $1.2 < x < 1.3$

Using **1**: there is a change of sign, so the equation $f(x) = 0$ has a root in the interval $1.2 < x < 1.3$.

Worked exam style question 1

(a) Show that the equation $6x - 3e^x + 4 = 0$ can be written in the form $x = \ln(2x + \frac{4}{3})$.

(b) Starting with $x_0 = 1.5$, use the iteration formula $x_{n+1} = \ln(2x_n + \frac{4}{3})$ to find x_3 to 2 decimal places.

(a) $6x - 3e^x + 4 = 0$

$$3e^x = 6x + 4$$

$$\frac{3e^x}{3} = \frac{6x}{3} + \frac{4}{3}$$

$$e^x = 2x + \frac{4}{3}$$

so $$x = \ln\left(2x + \frac{4}{3}\right)$$

Rearrange the equation by adding $3e^x$ to each side and dividing each term by 3

The inverse of e^x is $\ln x$

(b) $x_0 = 1.5$

$$x_1 = \ln\left(2(1.5) + \frac{4}{3}\right)$$

$$= 1.466337069$$

Using $x_{n+1} = \ln(2 \ln x_n + \frac{4}{3})$. Here $n = 0$ and $x_0 = 1.5$.

$$x_2 = \ln\left(2(1.466337069) + \frac{4}{3}\right)$$

$$= 1.450678371$$

Using $x_{n+1} = \ln(2 \ln x_n + \frac{4}{3})$. Here $n = 1$ and $x_1 = 1.466337069$.

$$x_3 = \ln\left(2(1.450678371) + \frac{4}{3}\right)$$

$$= 1.443310144$$

Using $x_{n+1} = \ln(2 \ln x_n + \frac{4}{3})$. Here $n = 2$ and $x_2 = 1.450678371$.

so $x_3 = 1.44$ to 2 decimal places

Worked exam style question 2

(a) Show that the equation $x^3 - 5x^2 - 7 = 0$ has a root close to $x = 5.25$.

(b) Rearrange the equation $x^3 - 5x^2 - 7 = 0$ in the form $x = \sqrt{\frac{p}{x + q}}$, where the values of p and q are to be found.

(c) Starting with $x_0 = 5.25$, show that the iteration formula $x_{n+1} = \sqrt{\frac{p}{x_n + q}}$, with the values of p and q found in part **(b)**, does not converge to a root of the equation.

(a) Let $f(x) = x^3 - 5x^2 - 7$

$f(5.24) = -0.410$

$f(5.25) = -0.109$

$f(5.26) = 0.194$

$f(5.25) < 0$ and $f(5.26) > 0$, so there is a root between $x = 5.25$ and $x = 5.26$

Show that the equation $f(x) = 0$ has a root in an interval near to $x = 5.25$

Look for a change of sign

Substitute $x = 5.24$, $x = 5.25$ and $x = 5.26$ into the function

Using **1**: there is a change of sign, so the equation $f(x) = 0$ has a root in the interval $5.25 < x < 5.26$.

(b)

$$x^3 - 5x^2 - 7 = 0$$

$$x^2(x - 5) - 7 = 0$$

$$x^2(x - 5) = 7$$

$$\frac{x^2(x - 5)}{(x - 5)} = \frac{7}{x - 5}$$

$$x^2 = \frac{7}{x - 5}$$

so $$x = \sqrt{\frac{7}{x - 5}}$$

Rearrange the equation by factorising $x^3 - 5x^2$, adding 7 to each side and dividing each side by $(x - 5)$.

Take the square root of each side and notice $p = 7$ and $q = -5$

(c) $x_0 = 5.25$

$$x_1 = \sqrt{\frac{7}{5.25 - 5}}$$

$$= 5.291502622$$

Using $x_{n+1} = \sqrt{\frac{7}{x_n - 5}}$. Here $n = 0$ and $x_0 = 5.25$.

$$x_2 = \sqrt{\frac{7}{5.291502622 - 5}}$$

$$= 4.900357755$$

Using $x_{n+1} = \sqrt{\frac{7}{x_n - 5}}$. Here $n = 1$ and $x_1 = 5.291502622$.

$$x_3 = \sqrt{\frac{7}{4.900357755 - 5}}$$

$$= \sqrt{-70.25132764}$$

Using $x_{n+1} = \sqrt{\frac{7}{x_n - 5}}$. Here $n = 2$ and $x_2 = 4.900357755$.

We havs a square root of a negative number, so the sequence $x_0, x_1, x_2, x_3 \ldots$ does not converge to a root.

Using **4**: an iteration formula does not necessarily give a sequence that converges to a root.

Revision exercise 4

1 The equation $x^3 - 4x^2 - 17x + 50 = 0$ has 3 real roots.
Find intervals in the form $(N, N + 1)$, where N is an integer, in which each root lies.

2 $f(x) = 2x^3 - 7x^2 + 2$

(a) Show that the equation $f(x) = 0$ has a root $x = a$, where a lies in the interval $3 < a < 4$.

(b) Show that the equation $2x^3 - 7x^2 + 2 = 0$ can be written in the form $x = 3.5 - \frac{1}{x^2}$.

3 $f(x) = x^4 - 3x + 1$

(a) Show that $f(x) = 0$ can be arranged in the form $x = \left(a - \frac{b}{x}\right)^{\frac{1}{3}}$,

where the values of a and b are to be found.

(b) Use the iteration formula $x_{n+1} = \left(a - \frac{b}{x_n}\right)^{\frac{1}{3}}$ with $x_0 = 1$ and your values of a and b to find the approximate solution x_4 of the equation, to an appropriate degree of accuracy.

4 **(a)** On the same axes, sketch the graphs of $y = \sin x$ and $y = x^2$. Hence write down the number of roots of the equation $\sin x = x^2$.

(b) Show that the equation $\sin x = x^2$ has a root in the interval $0.8 < x < 0.9$.

(c) Use the iteration formula $x_{n+1} = \frac{\sin x_n}{x_n}$ with $x_0 = 0.9$ to find in turn x_1, x_2, x_3 and x_4 giving your final answer to 3 decimal places.

5 $f(x) = e^x - x^3 + 1$

(a) Show that the equation $f(x) = 0$ has a root near to $x = 4.5$.

(b) Show that the equation $e^x - x^3 + 1 = 0$ can be written in the form $x = \ln(x^3 - 1)$.

(c) Starting with $x_0 = 4.5$, use the iteration formula $x_{n+1} = \ln(x_n^3 - 1)$ to find x_4 to 4 significant figures.

6 **(a)** By sketching the curves with equations $y = e^x + 1$ and $y = \frac{1}{x}$, show that the equation $e^x + 1 = \frac{1}{x}$ has only one root.

(b) Show that the root of the equation $e^x + 1 = \frac{1}{x}$ lies in the interval $0.4 < x < 0.41$.

(c) Starting with $x_0 = 0.4$, use the iteration formula $x_{n+1} = (e^{x_n} + 1)^{-1}$ to calculate this root to 3 decimal places.

7 **(a)** Use the iteration formula $x_{n+1} = \frac{2}{x_n^2 + 4}$, with $x_0 = 0.5$, to find x_4 to 3 significant figures.

The only root of the equation $x^3 + 4x - 2 = 0$ is α. It is given that, to 4 significant figures, $\alpha = x_4$.

(b) Use the substitution $y = 2^x$ to express $8^x + 2^{x+2} - 2 = 0$ as a cubic equation.

(c) Hence find an approximate solution to the equation $8^x + 2^{x+2} - 2 = 0$, giving your answer to 2 significant figures.

8 $f(x) = 2\tan x - x - 4$, where x is in radians.

(a) Evaluate f(1.2) and f(1.21) and deduce that there is a root α in the interval $1.2 < \alpha < 1.21$.

(b) Show that the equation $f(x) = 0$ can be arranged in the form $x = \tan^{-1}(ax + b)$, where the values of a and b are to be found.

(c) Starting with $x_0 = 1.2$, use the iteration formula $x_{n+1} = \tan^{-1}(ax_n + b)$, with your values of a and b, to calculate α to 3 decimal places.

9 $f(x) = \frac{x}{2} - (\sin x + \cos x)^2$

(a) Find the exact values of $f\left(\frac{\pi}{2}\right)$ and $f(\pi)$ and deduce that there is a root in the interval $\frac{\pi}{2} < x < \pi$.

An attempt to evaluate this root is made using the iteration formula $x_{n+1} = 2(\sin x_n + \cos x_n)^2$ with $x_0 = \pi$.

(b) Describe the result of such an attempt.

10 $f(x) = x^3 - x^{\frac{1}{3}} - 3$

(a) Copy and complete the table below.

x	−3	−2	−1	0	1	2	3
f(x)							

Given that the root of the equation $f(x) = 0$ lies in the interval $\alpha < x < \alpha + 1$, where α is an integer:

(b) write down the value of α,

(c) use the iteration formula $x_{n+1} = (x_n^{\frac{1}{3}} + 3)^{\frac{1}{3}}$ to find the root of the equation $f(x) = 0$ to 2 decimal places.

11 $f(x) = x^3 - 2x - 5$

(a) Show that there is a root α of $f(x) = 0$ for x in the interval $[2, 3]$.

The root α is to be estimated using the iterative formula

$$x_{n+1} = \left(2 + \frac{5}{x_n}\right)^{\frac{1}{2}}, \ x_0 = 2.$$

(b) Calculate the values of x_1, x_2, x_3 and x_4, giving your answers to 4 significant figures.

(c) Prove that, to 5 significant figures, α is 2.0946. *(E)*

12 $f(x) = 9e^{0.5x} - x^3$

(a) Find $f'(x)$.

(b) By evaluating $f'(2)$ and $f'(2.1)$, show that the curve with equation $y = f(x)$ has a stationary point at $x = q$, where $2 < q < 2.1$.

(c) Starting with $x_0 = 2$, use the iteration formula $x_{n+1} = \sqrt{1.5e^{0.5x_n}}$ to calculate q to 2 decimal places.

13

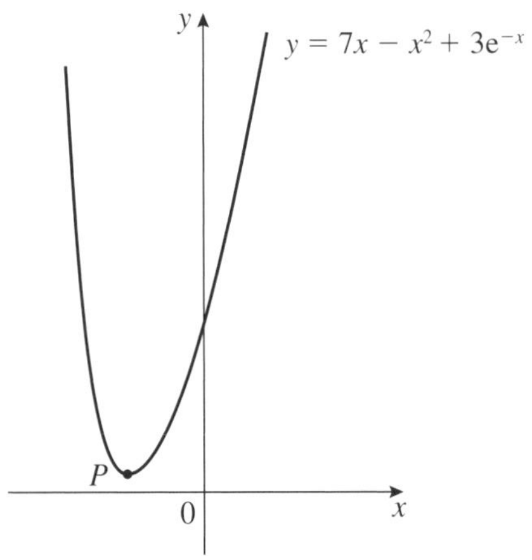

The diagram shows part of the curve with equation $y = f(x)$, where $f(x) = 7x - x^2 + 3e^{-x}$. The point P is the minimum point of the curve.

(a) Find $f'(x)$.

(b) Hence show that the x-coordinate of P is the solution of the equation $x = g(x)$, where $g(x) = -\ln[\frac{1}{3}(7 - 2x)]$.

(c) Starting with $x_0 = -1.1$, use the iteration formula $x_{n+1} = -\ln[\frac{1}{3}(7 - 2x_n)]$ to calculate x_1, x_2, x_3 and x_4. Hence write down the x-coordinate of P to 4 significant figures.

(d) Calculate the y-coordinate of P to 3 significant figures.

14 $f(x) = x^3 + x^2 - 4x - 1$

The equation $f(x) = 0$ has only one positive root, α.

(a) Show that $f(x) = 0$ can be rearranged as $x = \sqrt{\left(\frac{4x+1}{x+1}\right)}$, $x \neq -1$.

The iterative formula $x_{n+1} = \sqrt{\left(\frac{4x_n+1}{x_n+1}\right)}$ is used to find an approximation to α.

(b) Taking $x_1 = 1$, find to 2 decimal places, the values of x_2, x_3 and x_4.

(c) By choosing values of x in a suitable interval, prove that $\alpha = 1.70$, *correct* to 2 decimal places.

(d) Write down a value of x_1 for which the iteration formula $x_{n+1} = \sqrt{\left(\frac{4x_n+1}{x_n+1}\right)}$ does not produce a valid value for x_2.

Justify your answer. **E**

Test yourself	**What to review**
	If your answer is incorrect
1 Show that the equation $x^{\frac{1}{3}} - \sin x - 1 = 0$ has a root between $x = 2.73$ and $x = 2.74$.	*Review Heinemann Book C3 pages 43* *Revise for C3 page 26 Example 1*
2 Using the same axes, sketch the graphs of $y = \cos x$ and $y = \ln x$. Hence write down the number of roots of the equation $\ln x = \cos x$.	*Review Heinemann Book C3 pages 43* *Revise for C3 page 27 Example 2a*
3 Show that the equation $x^3 - 6x^2 + 3x - 6 = 0$ can be written in the form $x = \frac{6(x^2 + A)}{(x^2 + B)}$, where the values of A and B are to be found.	*Review Heinemann Book C3 page 47* *Revise for C3 page 28 Worked exam style question 2b*
4 Starting with $x_0 = 1$, use the iteration formula $x_{n+1} = \ln(\ln x_n + 5)$, to find, to 2 decimal places, the value of x_1, x_2, x_3 and x_4.	*Review Heinemann Book C3 page 49* *Revise for C3 page 27 Worked exam style question 1b*

Test yourself answers

1 (a) $f(2.73) = -0.002$, $f(2.74) = 0.008$

2 One root

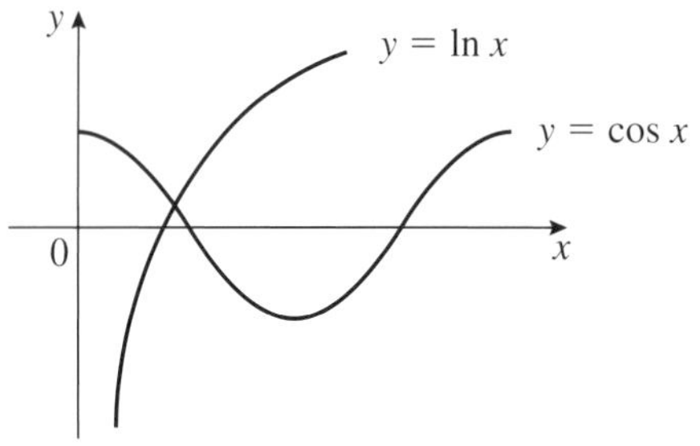

3 $A = 1$, $B = 3$

4 1.61, 1.70, 1.71, 1.71

Transforming graphs of functions

5

Key points to remember

1 The modulus of a number a, written as $|a|$, is its **positive** numerical value.

- For $|a| \geqslant 0$, $|a| = a$.
- For $|a| < 0$, $|a| = -a$.

2 To sketch the graph of $y = |f(x)|$:

- Sketch the graph of $y = f(x)$.
- Reflect in the x-axis any parts where $f(x) < 0$ (parts below the x-axis).
- Delete the parts below the x-axis.

3 To sketch the graph of $y = f(|x|)$:

- Sketch the graph of $y = f(x)$ for $x \geqslant 0$.
- Reflect this in the y-axis.

4 To solve an equation of the type $|f(x)| = g(x)$ or $|f(x)| = |g(x)|$:

- Use a sketch to locate the roots.
- Solve algebraically, using $-f(x)$ for reflected parts of $y = f(x)$ and $-g(x)$ for reflected parts of $y = g(x)$.

5 Basic types of transformation are:

$f(x + a)$	a horizontal translation of $-a$,
$f(x) + a$	a vertical translation of $+a$,
$f(ax)$	a horizontal stretch of scale factor $\frac{1}{a}$,
$af(x)$	a vertical stretch of scale factor a.

These may be combined to give, for example, $bf(x + a)$, which is a horizontal translation of $-a$ followed by a vertical stretch of scale factor b.

6 For combinations of transformations, the graph can be built up 'one step at a time', starting from a basic or given curve.

Example 1

Sketch the graph of $y = \left|\frac{1}{2}x + 5\right|$, showing the coordinates of any points at which the graph meets the axes.

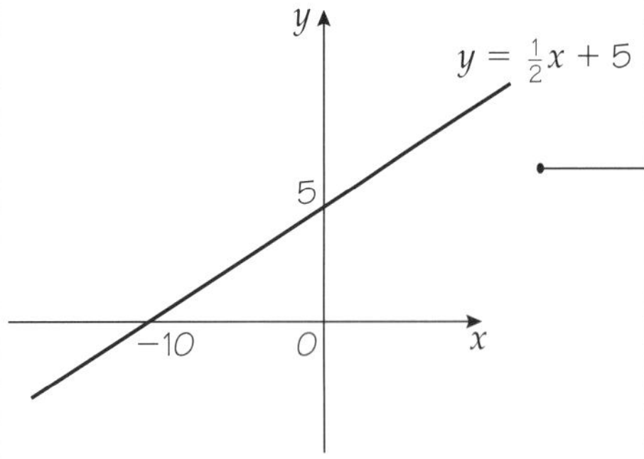

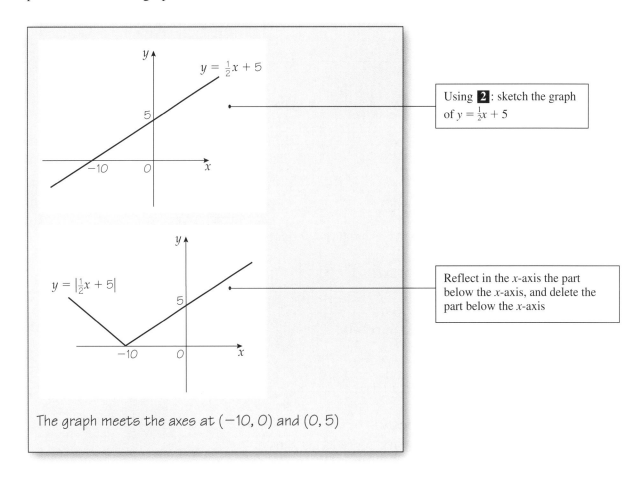

Using **2**: sketch the graph of $y = \frac{1}{2}x + 5$

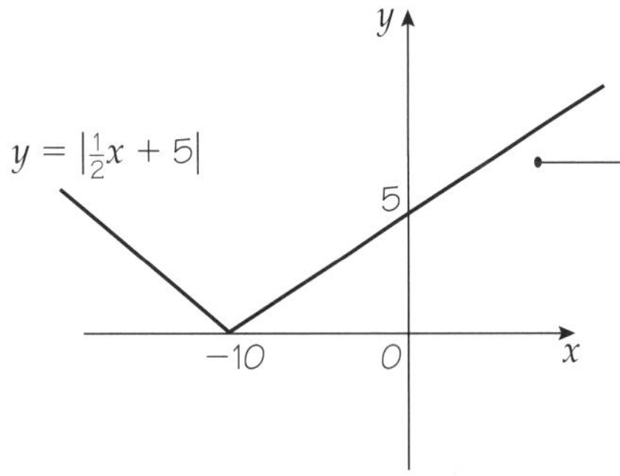

Reflect in the x-axis the part below the x-axis, and delete the part below the x-axis

The graph meets the axes at $(-10, 0)$ and $(0, 5)$

Example 2

Sketch the graph of $y = e^{|x|} - 2$.

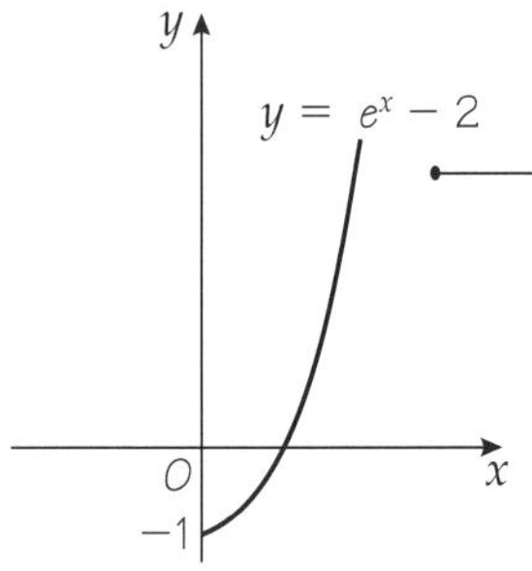

Using **3**: sketch the graph of $y = e^x - 2$, for $x \geqslant 0$ (use a vertical translation of -2 on the graph of $y = e^x$)

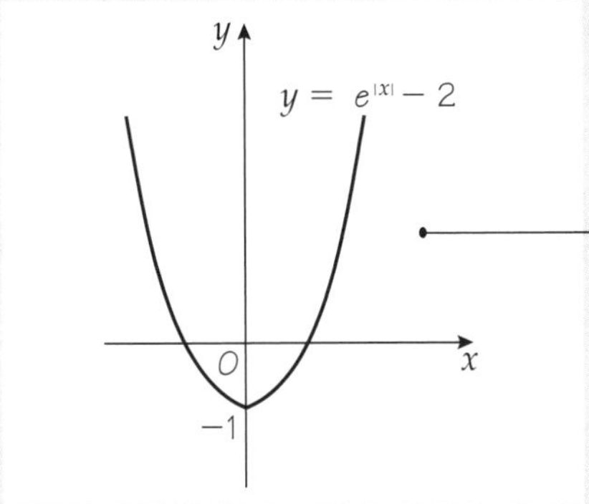

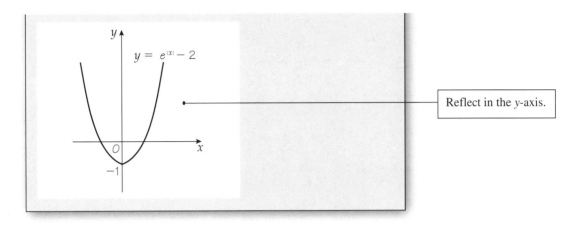

Reflect in the y-axis.

Example 3

Using combinations of transformations, sketch the graph of:

(a) $y = 1 + \dfrac{1}{x-2}$

(b) $y = 3 \sin 2x$, $0 \leqslant x \leqslant 360°$.

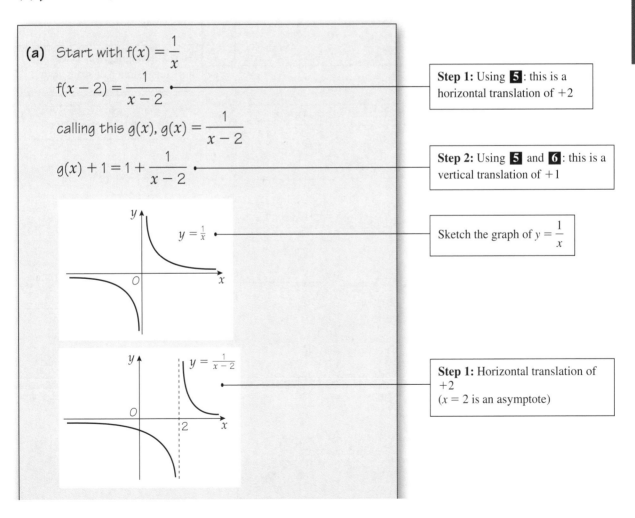

(a) Start with $f(x) = \dfrac{1}{x}$

$f(x-2) = \dfrac{1}{x-2}$

Step 1: Using **5**: this is a horizontal translation of $+2$

calling this $g(x)$, $g(x) = \dfrac{1}{x-2}$

$g(x) + 1 = 1 + \dfrac{1}{x-2}$

Step 2: Using **5** and **6**: this is a vertical translation of $+1$

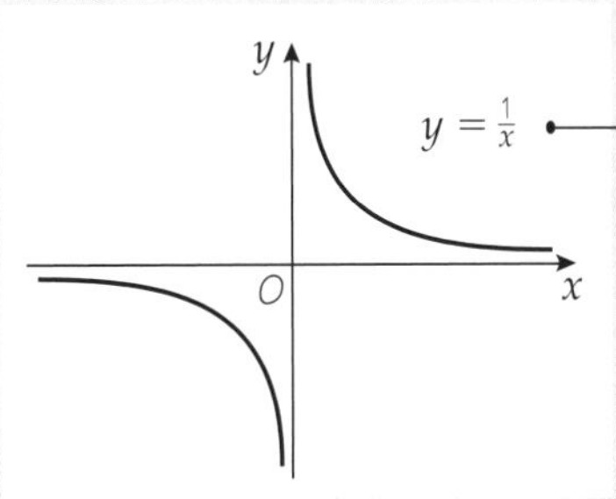

Sketch the graph of $y = \dfrac{1}{x}$

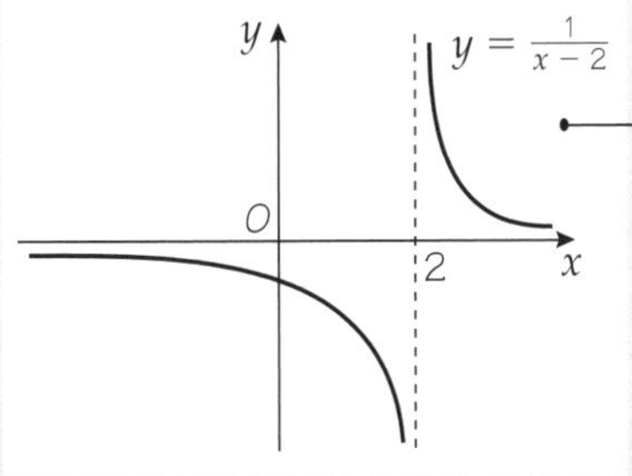

Step 1: Horizontal translation of $+2$
($x = 2$ is an asymptote)

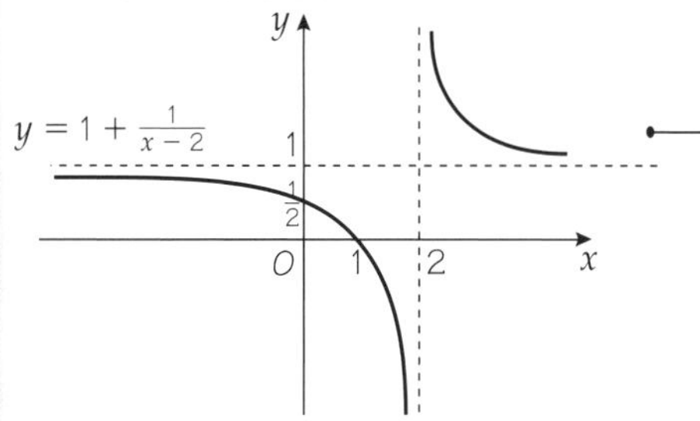

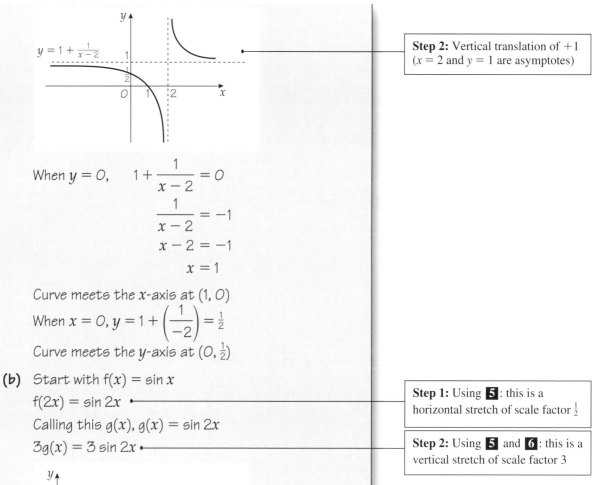

Step 2: Vertical translation of +1 ($x = 2$ and $y = 1$ are asymptotes)

When $y = 0$, $1 + \dfrac{1}{x-2} = 0$

$$\frac{1}{x-2} = -1$$

$$x - 2 = -1$$

$$x = 1$$

Curve meets the x-axis at (1, 0)

When $x = 0$, $y = 1 + \left(\dfrac{1}{-2}\right) = \frac{1}{2}$

Curve meets the y-axis at $(0, \frac{1}{2})$

(b) Start with $f(x) = \sin x$

$f(2x) = \sin 2x$

Step 1: Using **5**: this is a horizontal stretch of scale factor $\frac{1}{2}$

Calling this $g(x)$, $g(x) = \sin 2x$

$3g(x) = 3 \sin 2x$

Step 2: Using **5** and **6**: this is a vertical stretch of scale factor 3

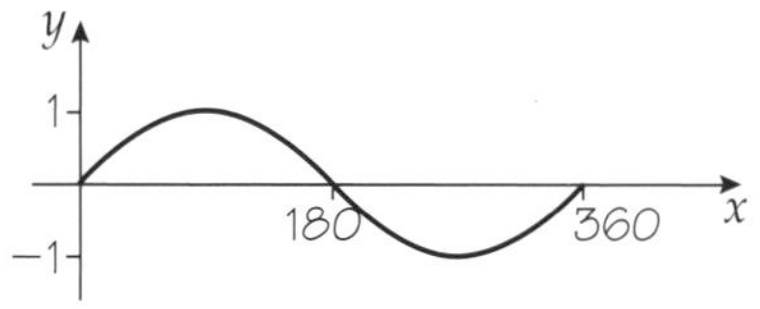

Sketch the graph of $y = \sin x$

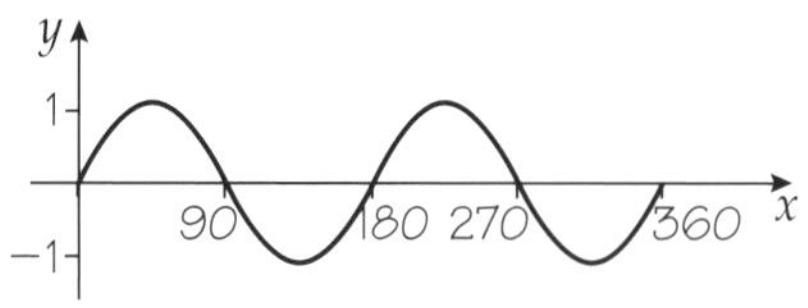

Step 1: Horizontal stretch of scale factor $\frac{1}{2}$

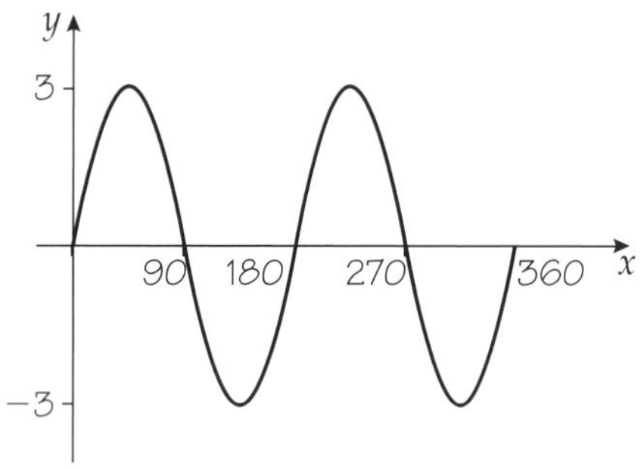

Step 2: Vertical stretch of scale factor 3

Worked exam style question 1

(a) On the same diagram, sketch the graph of $y = |x^2 - x - 2|$ and the graph of $y = 2x$.

(b) Solve the equation $|x^2 - x - 2| = 2x$, giving your answers to 2 decimal places where appropriate.

(a) For $y = x^2 - x - 2$

When $y = 0$, $x^2 - x - 2 = 0$

$$(x + 1)(x - 2) = 0$$

$$x = -1, x = 2$$

When $x = 0$, $y = -2$

> To help in sketching the graph, find the points at which it meets the axes

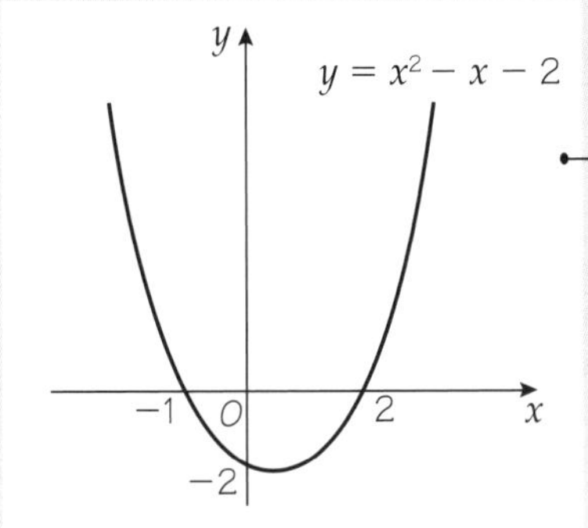

> Using **2**: sketch the graph of $y = x^2 - x - 2$

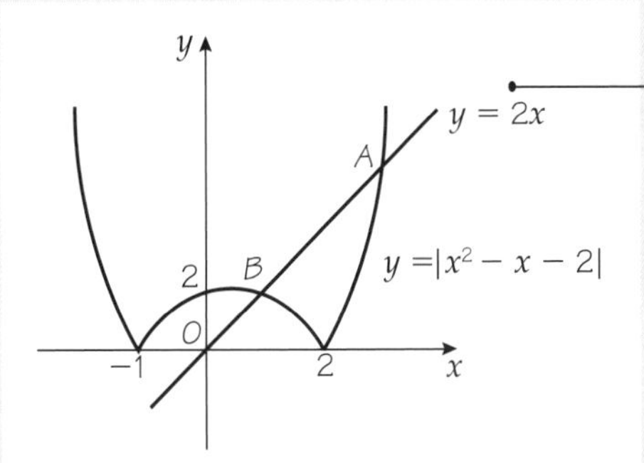

> Reflect in the x-axis the part below the x-axis, and delete the part below the x-axis
>
> Sketch the graph of $y = 2x$

(b) Solving $|x^2 - x - 2| = 2x$:

$$x^2 - x - 2 = 2x$$

> Using **4**: intersection point A is on the unreflected part of the curve

$$x^2 - 3x - 2 = 0$$

$$x = \frac{-(-3) \pm \sqrt{(-3)^2 - 4(-2)}}{2}$$

$$x = \frac{3 \pm \sqrt{17}}{2}$$

At A, $x = \dfrac{3 + \sqrt{17}}{2}$

> $x > 0$ at A, so the other solution is not valid

$$x = 3.56 \text{ (2 d.p.)}$$

$-(x^2 - x - 2) = 2x$

$-x^2 + x + 2 = 2x$

$-x^2 - x + 2 = 0$

$x^2 + x - 2 = 0$

$(x + 2)(x - 1) = 0$

$x = -2, x = 1$

At B, $x = 1$

Solutions are $x = 1$, $x = 3.56$

Intersection point B is on the reflected part of the curve. For the reflected part of $y = \mathrm{f}(x)$, use $-\mathrm{f}(x)$ to solve the equation.

$x > 0$ at B, so the other solution is not valid

Worked exam style question 2

The diagram shows part of the curve with equation $y = \mathrm{f}(x)$.

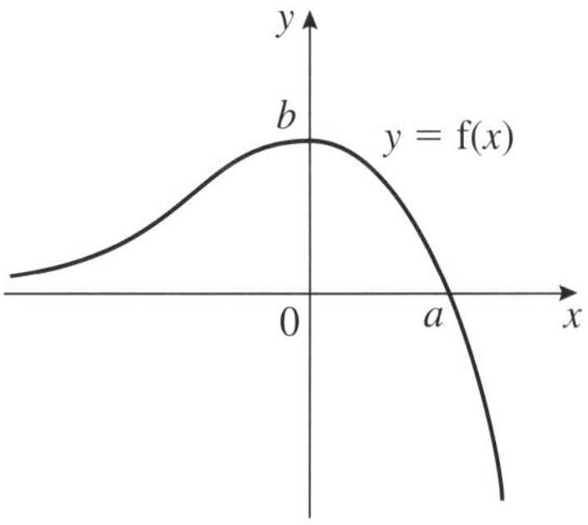

The curve meets the axes at the points with coordinates $(a, 0)$ and $(0, b)$, where a and b are positive constants.

On separate diagrams, sketch the curve with equation:

(a) $y = |\mathrm{f}(x)| + 2$

(b) $y = -3\mathrm{f}(4x)$.

In each case show, in terms of a or b, the coordinates of points at which the curve meets the axes.

(a)

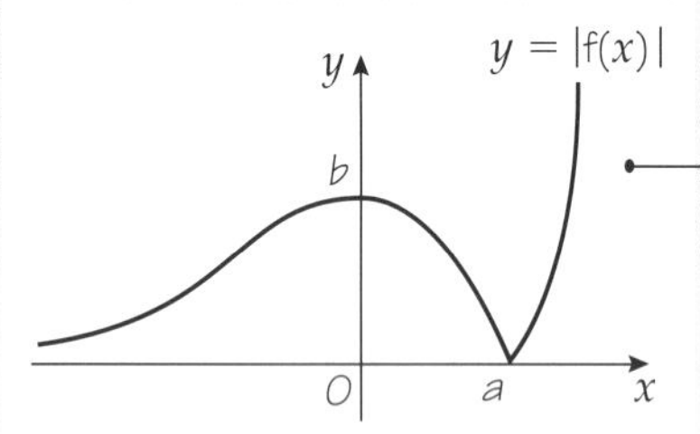

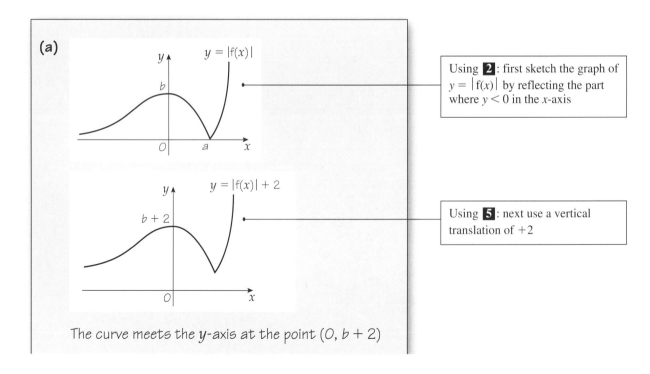

Using **2**: first sketch the graph of $y = |\mathrm{f}(x)|$ by reflecting the part where $y < 0$ in the x-axis

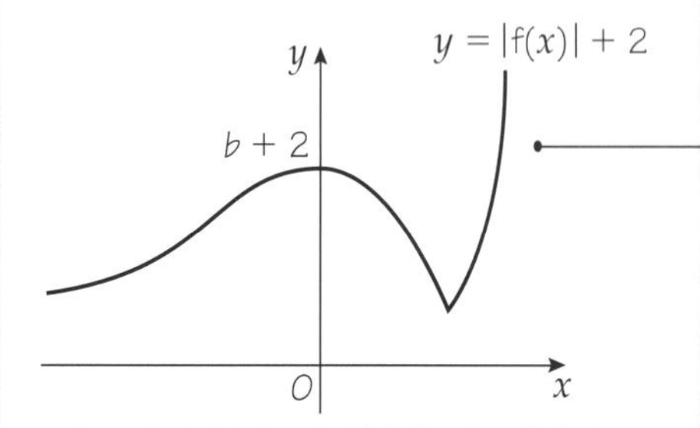

Using **5**: next use a vertical translation of $+2$

The curve meets the y-axis at the point $(0, b + 2)$

(b)

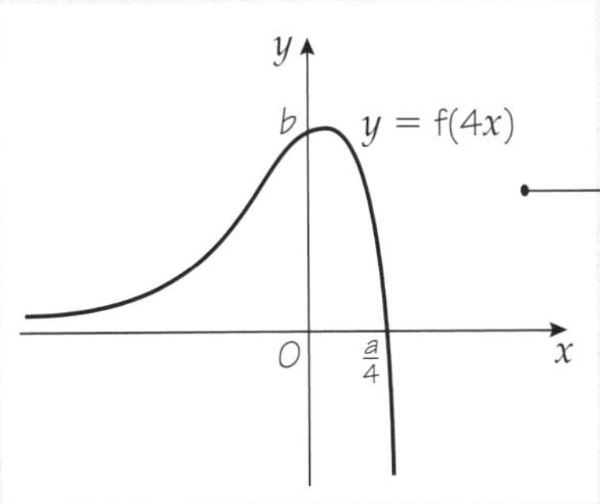

Using **5** and **6**

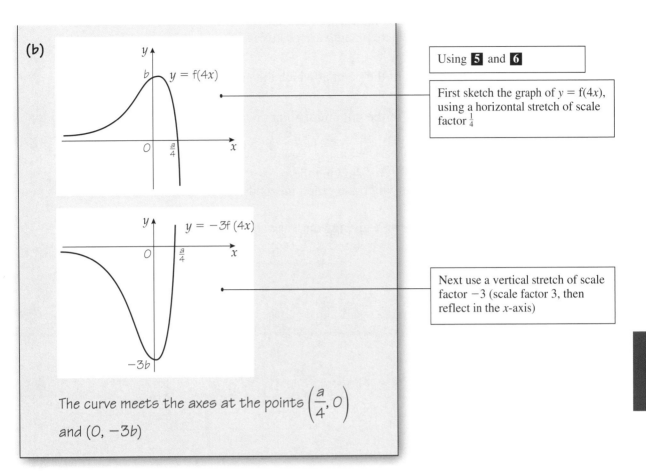

First sketch the graph of $y = f(4x)$, using a horizontal stretch of scale factor $\frac{1}{4}$

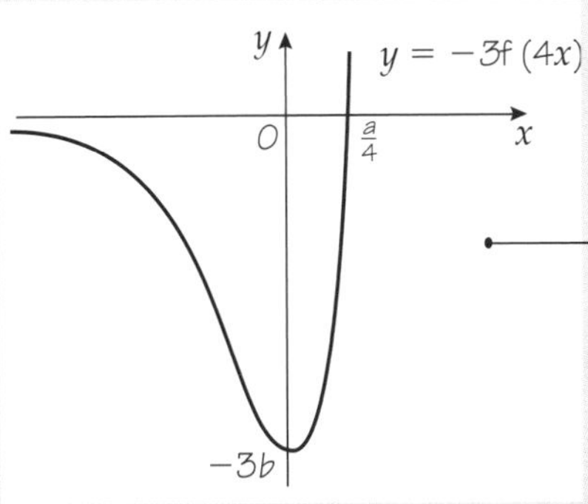

Next use a vertical stretch of scale factor -3 (scale factor 3, then reflect in the x-axis)

The curve meets the axes at the points $\left(\frac{a}{4}, 0\right)$ and $(0, -3b)$

Revision exercise 5

1 Sketch the graph of $y = |2x + 4|$, showing the coordinates of any points at which the graph meets the axes.

2 Sketch the graph of $y = |x^2 - 4x + 3|$, showing the coordinates of any points at which the graph meets the axes.

3 Sketch the graph of $y = 3|x| - 2$, showing the coordinates of any points at which the graph meets the axes.

4 **(a)** On the same diagram, sketch the graph of $y = |x - 2|$ and the graph of $y = \frac{1}{3}x + 1$.

(b) Solve the equation $|x - 2| = \frac{1}{3}x + 1$.

5 **(a)** On the same diagram, for $0 \leqslant x \leqslant 2\pi$, sketch the graph of $y = |\sin x|$ and the graph of $y = |\cos x|$.

(b) Find the exact coordinates of the points of intersection of the graphs in part **(a)**.

6 (a) Using a combination of transformations, sketch the graph of $y = (x + p)^2 + q$, where p and q are positive constants.

(b) Write down the coordinates of the point at which the graph crosses the y-axis.

(c) Write down the coordinates of the minimum point on the graph.

7 (a) Sketch the graph of $y = |5^x - 5|$, showing the coordinates of any points at which the graph meets the axes.

(b) Solve the equation $|5^x - 5| = 2$, giving your answers to 2 decimal places.

8

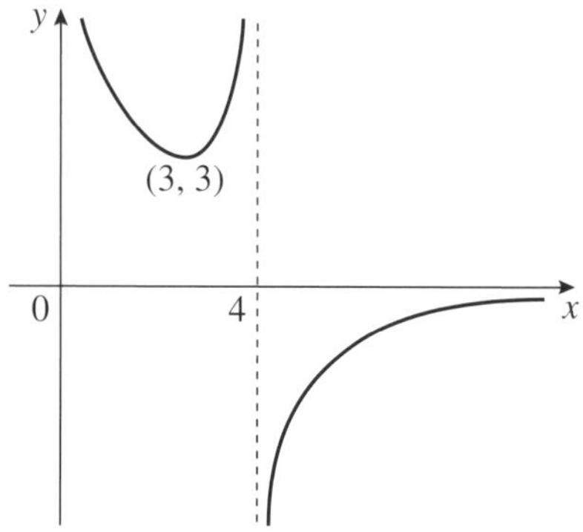

The diagram shows a sketch of part of the curve with equation $y = f(x)$, $x > 0$.

The curve has a minimum point at (3, 3), and the lines $x = 4$, the x-axis and the y-axis are asymptotes of the curve.

(a) Sketch the graph of $y = |f(x) - 2|$, $x > 0$.

(b) Write down the coordinates of the turning point of this curve.

(c) Write down the equations of the asymptotes of this curve.

9 The function f is defined by $f: x \rightarrow |2x - 5|$, $x \in \mathbb{R}$.

(a) Sketch the graph of $y = f(x)$, showing the coordinates of points at which the graph meets or crosses the axes.

(b) Find the values of x for which $f(x) = x$.

The function g is defined by $g: x \rightarrow x(x - 6)$, $x \in \mathbb{R}$.

(c) Find the range of $g(x)$.

(d) Find fg(1). (E)

10 The diagram shows a sketch of the curve with equation $y = f(x), x \geqslant 0$. The curve meets the coordinate axes at the points $(0, c)$ and $(d, 0)$.

In separate diagrams, sketch the curve with equation

(a) $y = f^{-1}(x)$

(b) $y = 3f(2x)$.

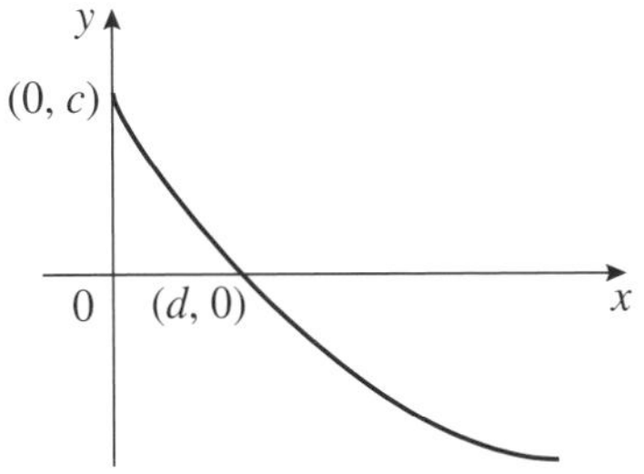

Indicate clearly on each sketch the coordinates, in terms of c or d, of any point where the curve meets the coordinate axes.

Given that f is defined by $f : x \rightarrow 3(2^{-x}) - 1, x \in \mathbb{R}, x \geqslant 0$:

(c) state **(i)** the value of c, **(ii)** the range of f.

(d) Find the value of d, giving your answer to 3 decimal places.

The function g is defined by $g : x \rightarrow \log_2 x, x \in \mathbb{R}, x \geqslant 1$.

(e) Find $fg(x)$, giving your answer in its simplest form.

E

Test yourself	**What to review**
	If your answer is incorrect
1 Sketch the graph of $y = \lvert \frac{1}{4}x - 1 \rvert$, showing the coordinates of any points at which the graph meets the axes.	*Review Heinemann Book C3 pages 54–56* *Revise for C3 page 36 Example 1*
2 Sketch the graph of $y = \lvert x \rvert^2 - 3\lvert x \rvert$, showing the coordinates of any points at which the graph meets the axes.	*Review Heinemann Book C3 pages 57–58* *Revise for C3 page 36 Example 2*
3 **(a)** On the same diagram, sketch the graph of $y = \lvert 2x \rvert$ and the graph of $y = \lvert x - 5 \rvert$. **(b)** Solve the equation $\lvert 2x \rvert = \lvert x - 5 \rvert$.	*Review Heinemann Book C3 page 59–61* *Revise for C3 page 39 Worked exam style question 1*

4 Using a combination of transformations, sketch the graph of $y = \frac{1}{2}x^3 - 2$.

Review Heinemann Book C3 page 62–66
Revise for C3 page 37 Example 3

5

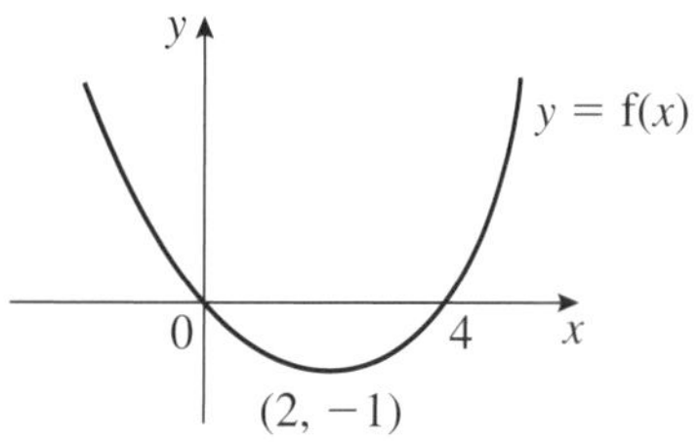

Review Heinemann Book C3 page 67–69
Revise for C3 page 40 Worked exam style question 2

The diagram shows part of the curve with equation $y = \mathrm{f}(x)$. The curve passes through the origin and the point (4, 0) and has a minimum at the point (2, −1).

On separate diagrams, sketch the curve with equation:

(a) $y = 4\mathrm{f}(x + 1)$

(b) $y = -\mathrm{f}(\frac{1}{2}x)$.

In each case, show the coordinates of any points at which the curve meets the x-axis and of the turning point of the curve.

---

6 **(a)** On the same diagram, sketch the graph of $y = \left|\frac{1}{x}\right|$ and the graph of $y = |3x + 2|$.

(b) Using algebra, find the coordinates of the points at which the two graphs intersect.

Review Heinemann Book C3 pages 59–61
Revise for C3 page 39
Work exam style question 1

Test yourself answers

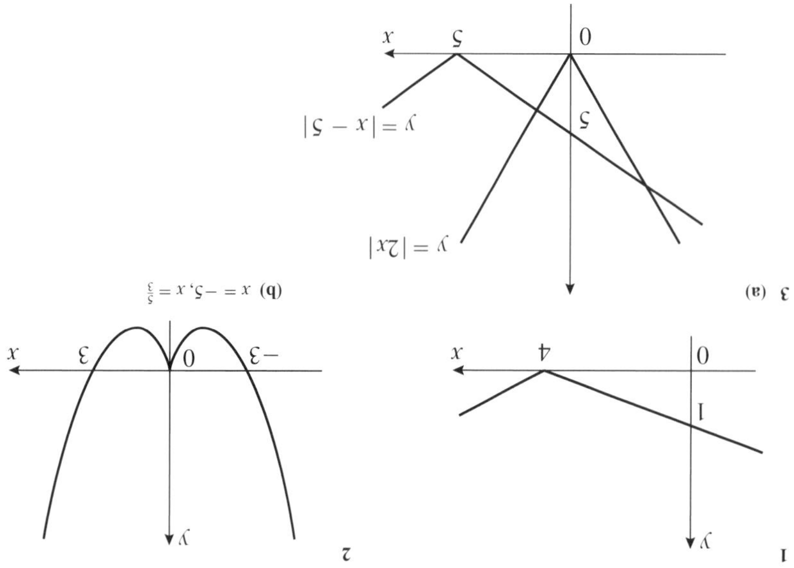

4

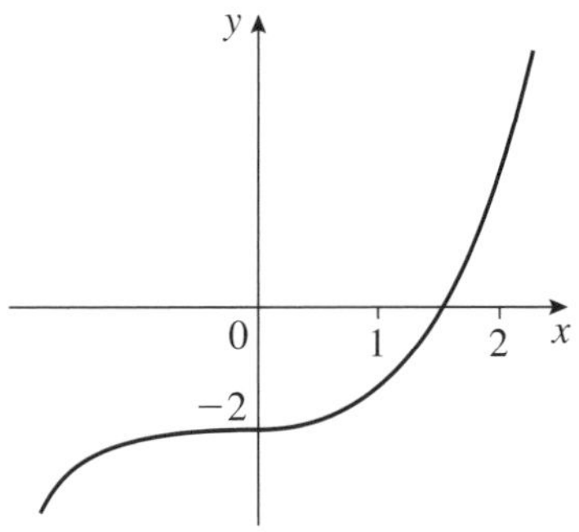

5 **(a)**

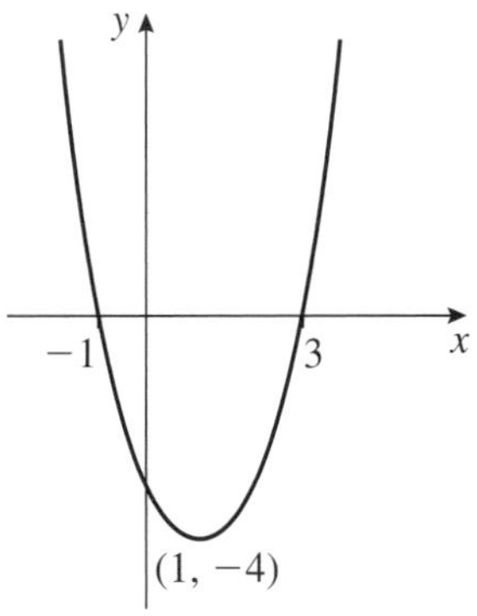

(−1, 0) and (3, 0)
Minimum at (1, −4)

(b)

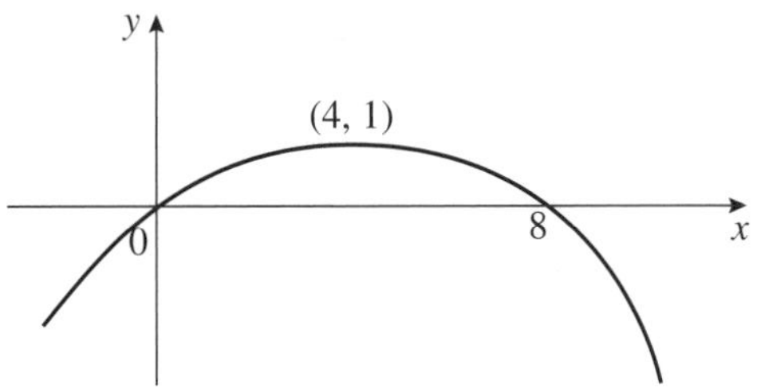

(0, 0) and (8, 0)
Maximum at (4, 1)

6 **(a)**

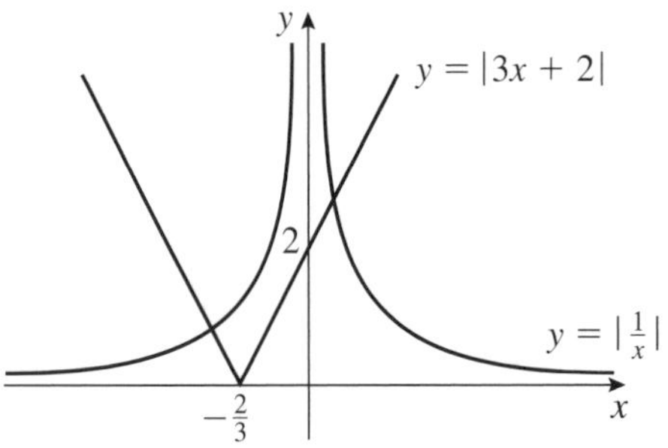

(b) (−1, 1) and $(\frac{1}{3}, 3)$

More trigonometric functions and related identities

6

Key points to remember

(When working in degrees replace π by 180°)

1 $\sec\theta = \dfrac{1}{\cos\theta}$

{$\sec\theta$ is undefined when $\cos\theta = 0$, i.e. at $\theta = (2n+1)\dfrac{\pi}{2}, n \in Z$}

The graph of $\sec\theta$ is:

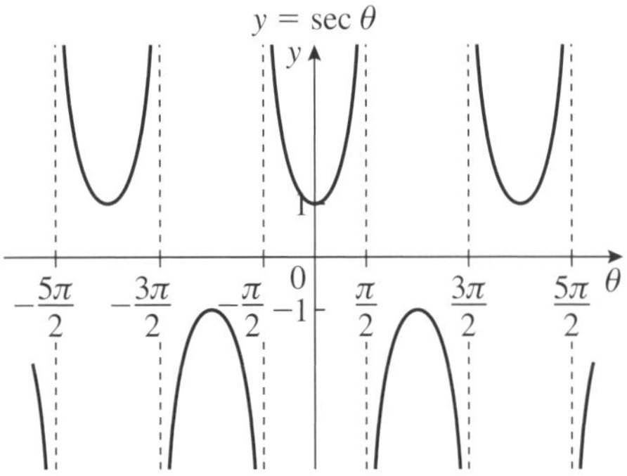

2 $\text{cosec}\,\theta = \dfrac{1}{\sin\theta}$

{$\text{cosec}\,\theta$ is undefined when $\sin\theta = 0$, i.e. at $\theta = \pi n°, n \in Z$}

The graph of $\text{cosec}\,\theta$ is:

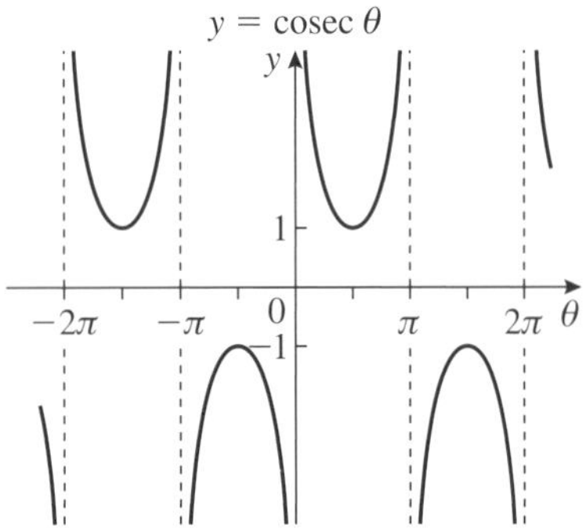

3 $\cot\theta = \dfrac{1}{\tan\theta}$

{$\cot\theta$ is undefined when $\tan\theta = 0$, i.e. at $\theta = \pi n°, n \in Z$}

($\cot\theta$ can also be written as $\dfrac{\cos\theta}{\sin\theta}$)

The graph of $\cot\theta$ is:

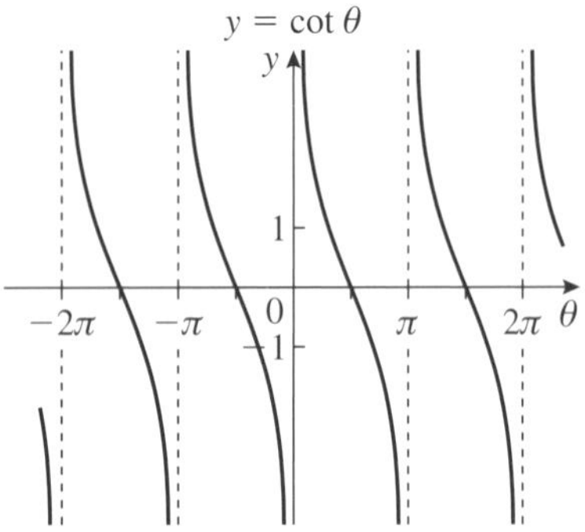

4 Two further Pythagorean identities, derived from $\sin^2\theta + \cos^2\theta \equiv 1$, are

(i) $1 + \tan^2\theta \equiv \sec^2\theta$ *and* (ii) $1 + \cot^2\theta \equiv \text{cosec}^2\theta$.

5 For $\sin x$, $\cos x$ and $\tan x$ to have inverse functions, $\arcsin x$ ($\sin^{-1} x$), $\arccos x$ ($\cos^{-1} x$) and $\arctan x$ ($\tan^{-1} x$) respectively, their domains need to be restricted so that they are one-to-one functions. The graphs of the inverse functions are those of the restricted functions reflected in the line $y = x$.

6 Restricting the domain of $\sin x$ to $-\frac{\pi}{2} \leqslant x \leqslant \frac{\pi}{2}$, $\arcsin x$ has domain $-1 \leqslant x \leqslant 1$ and range $-\frac{\pi}{2} \leqslant \arcsin x \leqslant \frac{\pi}{2}$.

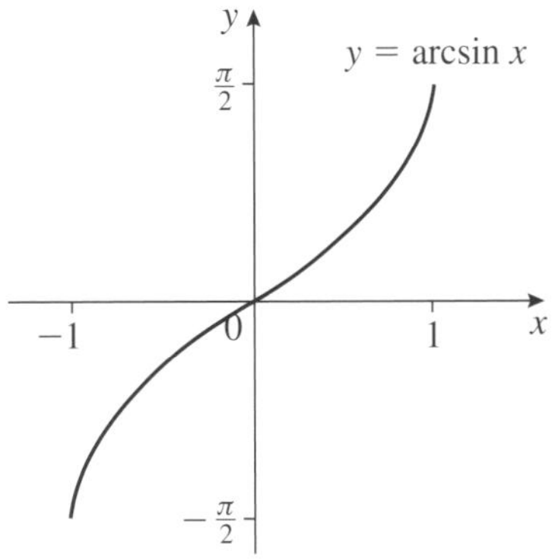

7 $\arcsin k$, $-1 \leqslant k \leqslant 1$, is the value of x, in the interval $-\frac{\pi}{2} \leqslant x \leqslant \frac{\pi}{2}$, for which $\sin x = k$.

8 Restricting the domain of $\cos x$ to $0 \leqslant x \leqslant \pi$, $\arccos x$ has domain $-1 \leqslant x \leqslant 1$, and range $0 \leqslant \arccos x \leqslant \pi$.

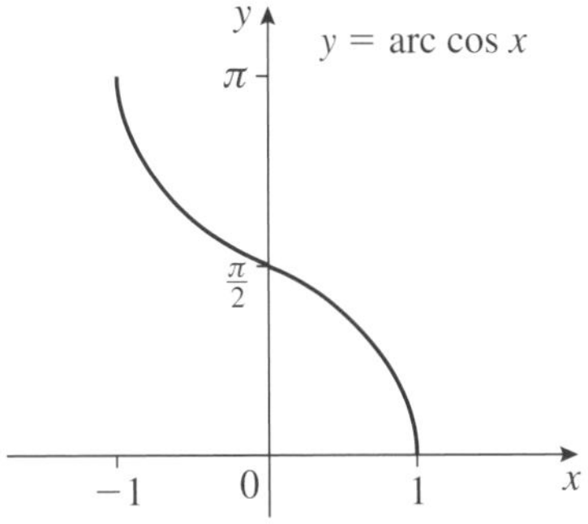

9 $\arccos k$, $-1 \leqslant k \leqslant 1$, is the value of x, in the interval $0 \leqslant x \leqslant \pi$, for which $\cos x = k$.

10 Restricting the domain of $\tan x$ to $-\frac{\pi}{2} < x < \frac{\pi}{2}$, $\arctan x$ has domain $x \in \mathbb{R}$, and range $-\frac{\pi}{2} < \arctan x < \frac{\pi}{2}$.

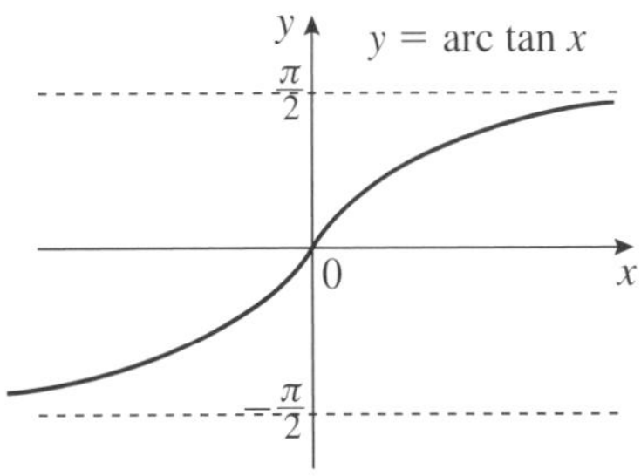

11 arctan k, $k \in \mathbb{R}$, is the value of x, in the interval $-\frac{\pi}{2} < x < \frac{\pi}{2}$, for which $\tan x = k$.
(In the examples and exercises arcsin x, arccos x and arctan x are understood to be as defined in **6**, **8** and **10** respectively.)

Example 1

Simplify:

(a) $\cos\theta(\operatorname{cosec}\theta + \tan\theta\sec\theta)$

(b) $(1 + \cot^2\theta)(1 - \cos^2\theta)$.

(a) $\cos\theta(\operatorname{cosec}\theta + \tan\theta\sec\theta) = \cos\theta\left(\frac{1}{\sin\theta} + \tan\theta\frac{1}{\cos\theta}\right)$ — Using **2** and **1**

$= \frac{\cos\theta}{\sin\theta} + \tan\theta\frac{\cos\theta}{\cos\theta}$ — Multiplying out

$= \cot\theta + \tan\theta$ — Using **3**

(b) $(1 + \cot^2\theta)(1 - \cos^2\theta) = \operatorname{cosec}^2\theta(1 - \cos^2\theta)$ — Using **4**(ii)

$= \operatorname{cosec}^2\theta\sin^2\theta$ — Using $\sin^2\theta + \cos^2\theta \equiv 1$

$= \left(\frac{1}{\sin\theta}\right)^2\sin^2\theta$ — Using **2**

$= 1$ — As $\left(\frac{1}{\sin\theta}\right)^2 = \frac{1}{\sin^2\theta}$

Example 2

Solve the equation $\sec 2x = -2.5$, in the interval $0 \leqslant x \leqslant 360°$, giving your answer to 1 decimal place.

$\sec 2x = -2.5 \Rightarrow \cos 2x = -0.4$ — Using **1** and $\frac{1}{2.5} = 0.4$

so $2x = 113.6°, 246.4°, 473.6°, 606.4°$ — Remember that solutions are in the second and third quadrants (as cos is negative), in the interval $0 \leqslant 2x \leqslant 720°$

$x = 56.8°, 123.2°, 236.8°, 303.2°$

Example 3

Show that $\sin(\arctan \frac{1}{2}) = \frac{\sqrt{5}}{5}$.

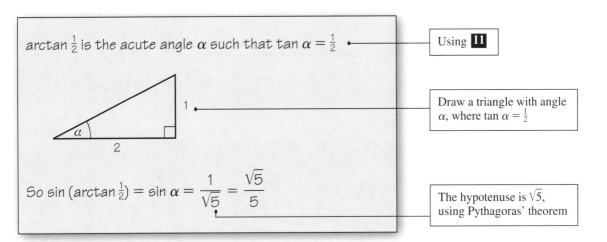

$\arctan \frac{1}{2}$ is the acute angle α such that $\tan \alpha = \frac{1}{2}$

Using **11**

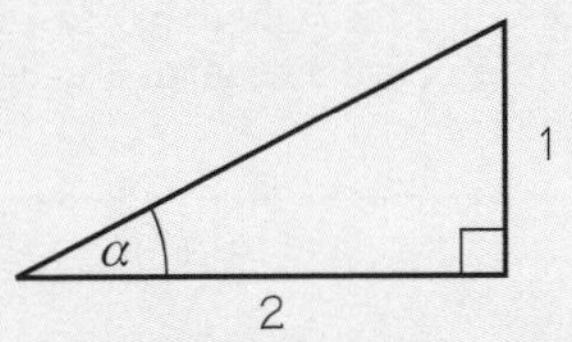

Draw a triangle with angle α, where $\tan \alpha = \frac{1}{2}$

So $\sin(\arctan \frac{1}{2}) = \sin \alpha = \frac{1}{\sqrt{5}} = \frac{\sqrt{5}}{5}$

The hypotenuse is $\sqrt{5}$, using Pythagoras' theorem

Worked exam style question 1

Show that $4 + (\tan\theta - \cot\theta)^2 \equiv \sec^2\theta + \operatorname{cosec}^2\theta$.

$$\text{LHS} \equiv 4 + \tan^2\theta - 2\tan\theta\cot\theta + \cot^2\theta$$

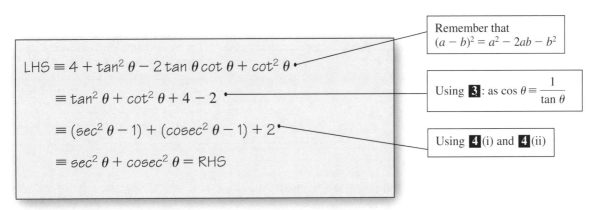

Remember that $(a - b)^2 = a^2 - 2ab - b^2$

$$\equiv \tan^2\theta + \cot^2\theta + 4 - 2$$

Using **3**: as $\cos\theta \equiv \frac{1}{\tan\theta}$

$$\equiv (\sec^2\theta - 1) + (\operatorname{cosec}^2\theta - 1) + 2$$

Using **4**(i) and **4**(ii)

$$\equiv \sec^2\theta + \operatorname{cosec}^2\theta = \text{RHS}$$

Worked exam style question 2

(a) Given that $2 \sin x \cos y - 5 \cos x \sin y = 3 \sin x \sin y$, express $\cot y$ in terms of $\cot x$.

(b) Given also that $\tan x = 5$, and that y is a reflex angle, find the exact value of $\operatorname{cosec} y$.

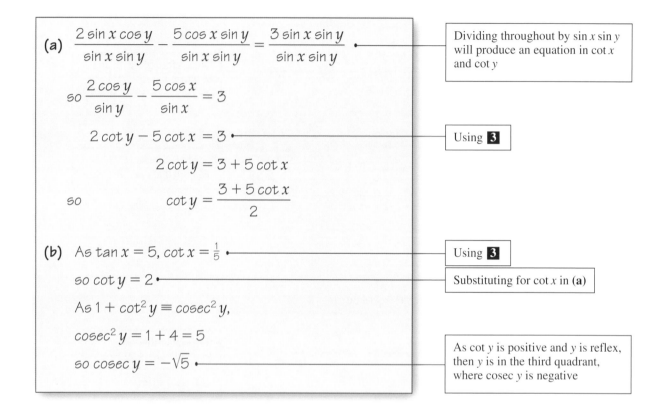

(a) $\dfrac{2 \sin x \cos y}{\sin x \sin y} - \dfrac{5 \cos x \sin y}{\sin x \sin y} = \dfrac{3 \sin x \sin y}{\sin x \sin y}$ — Dividing throughout by $\sin x \sin y$ will produce an equation in $\cot x$ and $\cot y$

so $\dfrac{2 \cos y}{\sin y} - \dfrac{5 \cos x}{\sin x} = 3$

$2 \cot y - 5 \cot x = 3$ — Using **3**

$2 \cot y = 3 + 5 \cot x$

so $\cot y = \dfrac{3 + 5 \cot x}{2}$

(b) As $\tan x = 5$, $\cot x = \frac{1}{5}$ — Using **3**

so $\cot y = 2$ — Substituting for $\cot x$ in **(a)**

As $1 + \cot^2 y \equiv \operatorname{cosec}^2 y$,

$\operatorname{cosec}^2 y = 1 + 4 = 5$

so $\operatorname{cosec} y = -\sqrt{5}$ — As $\cot y$ is positive and y is reflex, then y is in the third quadrant, where $\operatorname{cosec} y$ is negative

Worked exam style question 3

(a) Sketch on the same axes, in the interval $0 \leqslant x \leqslant 180°$, the graphs of $y = \cot x$ and $y = 2 \cos x$, showing the coordinates of any points of intersection with the axes.

(b) Find the coordinates of the points of intersection of the two curves in the given interval.

(a)

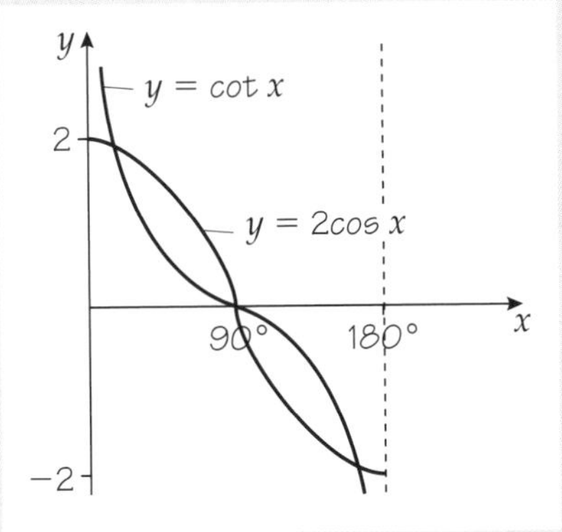

(b) As $\cot x = 2\cos x$ — The curves meet where $\cot x = 2\cos x$

$\dfrac{\cos x}{\sin x} = 2\cos x$ — Using **3**

so $\cos x = 2\cos x \sin x$

$\cos x\,(1 - 2\sin x) = 0$ — Factorise. Do not cancel $\cos x$

so $\cos x = 0$ or $\sin x = \frac{1}{2}$

$\cos x = 0$ gives the point $(90, 0)$

$\sin x = \frac{1}{2}$ gives a 'first' value of x as 30

The corresponding y-coordinate is $2\cos 30° = \sqrt{3}$

The coordinates of the other point of intersection are $(150, -\sqrt{3})$ — The symmetry of the graphs can be used to show this

Revision exercise 6

1 Simplify the following expressions to a single term.

(a) $\left(\dfrac{1 - \sin^2\theta}{\sin^2\theta}\right)$ **(b)** $\cot 2\theta \sin 2\theta$ **(c)** $\dfrac{1}{\sec^2\theta - 1}$

(d) $\operatorname{cosec}\theta - \operatorname{cosec}\theta\cos^2\theta$

(e) $1 + 2\tan^2\theta + \tan^4\theta$

(f) $\sqrt{\operatorname{cosec}^2 4\theta - \cot^2 4\theta}$

2 Without using your calculator, find the value of:

(a) $\sec 180°$ **(b)** $\cot 30°$ **(c)** $\sec 315°$ **(d)** $\operatorname{cosec}\dfrac{7\pi}{6}$.

3 Find, in the interval $0 \leqslant x \leqslant 360°$, the solutions of the following equations, giving answers, where necessary, to 1 decimal place.

(a) $\cot x = -\sqrt{3}$ **(b)** $\sec 2x = 2$

(c) $\sec x = 2 \operatorname{cosec} x$ **(d)** $\sec^2 x = 1 + \tan x$

(e) $\cot (2x + 30°) = -1$ **(f)** $\operatorname{cosec} x = 3 \sin x$

(g) $\sqrt{1 + \tan^2 2x} = 1$

4 On separate sets of axes, sketch the intervals indicated in the graphs of the curves with the following equations. Show the coordinates of points of intersection with the axes, and any maximum and minimum points.

(a) $y = \sec \theta, \ -180° \leqslant \theta \leqslant 180°$

(b) $y = \cot 2\theta, \ 0 \leqslant \theta \leqslant 360°$

(c) $y = 2 \sec \theta - 1, \ -\pi \leqslant \theta \leqslant \pi$

(d) $y = -\operatorname{cosec} (\theta - 30°), \ 0 \leqslant \theta \leqslant 360°$

5 Prove the following identities.

(a) $\sin^2 \theta + \tan^2 \theta \equiv \sec^2 \theta - \cos^2 \theta$

(b) $\tan \theta + \cot \theta \equiv \operatorname{cosec} \theta \sec \theta$

(c) $\sec^4 \theta - \tan^2 \theta \equiv \tan^4 \theta + \sec^2 \theta$

(d) $\dfrac{\cos \theta}{\operatorname{cosec} \theta - 1} + \dfrac{\cos \theta}{\operatorname{cosec} \theta + 1} \equiv 2 \tan \theta$

6 Given that $\sin x = p$ and $\sin y = q$, find, in terms of p and/or q, the values of:

(a) $\operatorname{cosec} x$ **(b)** $\cot^2 y$ **(c)** $\dfrac{\tan x \cot y}{\sec x \cos y}$.

7 Without using your calculator, and using the definitions for $\arcsin x$, $\arccos x$ and $\arctan x$ given at the start of the chapter, find the value of:

(a) $\cos \{\arcsin (-1)\}$ **(b)** $\sin \{\arccos (-\tfrac{1}{2})\}$

(c) $\dfrac{\arcsin (\frac{1}{2})}{\arctan (\sqrt{3})}$.

8 Find the value of $\arccos \left(-\dfrac{\sqrt{3}}{2}\right) - \arcsin \left(-\dfrac{\sqrt{2}}{2}\right)$, giving your answer in radians in terms of π.

9 In triangle ABC, $AB = 5$ cm, $AC = 2$ cm and $BC = 6$ cm.

(a) Show that $\sec A = -\frac{20}{7}$.

Use the cosine rule to find $\cos A$

(b) Find the exact value of cosec B.

Use the sine rule to find $\sin B$

10 **(a)** Solve, in the interval $0 \leqslant \theta < \pi$, the equation $\sec^2 3\theta = 2\tan 3\theta$.

(b) Find all the values of x, in the interval $0 \leqslant x < 360°$, for which $\tan^2 x = 5\sec x - 2$. Give your answers correct to one decimal place.

11 **(a)** Solve, in the interval $-\pi \leqslant \theta < \pi$, the equation $\cot^2 \theta = \operatorname{cosec} \theta + 1$.

(b) Find all the values of x, in the interval $0 \leqslant x < 360°$, for which $\operatorname{cosec}^2 x = 3\cot 2x + \cot^2 x$. Give your answers correct to one decimal place.

12 Show that $\tan\{\arccos(\frac{1}{3})\} = 2\sqrt{2}$.

13 Given that $\sin B = \frac{3}{4}$, and that angle B is obtuse,

(a) write down the value of cosec B,

(b) find the exact value of sec B.

14 Given that $\sec A = 3$, and that angle A is reflex, find the exact value of:

(a) $\tan A$ **(b)** $\cot A$ **(c)** cosec A.

15 Given that $3\sin x \sin y + 4\cos x \cos y = 5\sin x \cos y$, express $\cot y$ in terms of $\cot x$.

16 **(a)** Find the value of $\arcsin(\frac{1}{2}) + \arccos(\frac{1}{2})$, giving your answer in radians in terms of π.

(b) Given that angles A and B are defined as

$$A = \arcsin x \text{ and } B = \arccos x,\ -1 \leqslant x \leqslant 1,$$

show that $\sin A = \cos B$.

(c) Hence deduce that $\arcsin x + \arccos x = \dfrac{\pi}{2}$.

17 **(a)** Factorise $ab - a - b + 1$.

(b) Solve, in the interval $0 \leqslant x \leqslant 2\pi$, the equation $\cot x \operatorname{cosec} x - \operatorname{cosec} x - \cot x + 1 = 0$.

18 **(a)** Solve the equation $\arcsin x = 1$, giving your answer in radians to 3 significant figures.

(b) Given that $y = \arcsin x$, $-1 \leqslant x \leqslant 1$, sketch the graph $y = \arcsin x - 1$, giving the coordinates of any points of intersection with the axes.

19 **(a)** On the same set of axes, in the interval $-\pi \leqslant \theta \leqslant \pi$, sketch the graphs of $y = \sin\theta$ and $y = \sec 2\theta - 1$.

(b) Find the θ-coordinates of all the points of intersection of the graphs in the given interval, giving answers to 3 significant figures where necessary.

In **(b)** you will need to use an appropriate double angle formula (see Chapter 7).

20 The graph of $y = \cot^2 x$ and $y = 1 - \operatorname{cosec} x$ intersect, in the interval $-180° < x < 180°$, at the points A and B. Find the x-coordinates of A and B.

Test yourself	What to review
	If your answer is incorrect
1 Simplify: **(a)** $\sec\theta\sqrt{\operatorname{cosec}^2\theta - 1}$ **(b)** $(\cot x \sec x - \cot x)(\sec x + 1)$.	*Review Heinemann Book C3 pages 73–84* *Revise for C3 page 48 Example 1*
2 Solve, in the interval $0 \leqslant \theta < 360°$, $\operatorname{cosec} 2\theta = -2$.	*Review Heinemann Book C3 pages 73–74* *Revise for C3 page 48 Example 2*
3 Given that $\cot x = -1.5$, and that x is obtuse, find the exact value of $\operatorname{cosec} x$.	*Review Heinemann Book C3 pages 73, 83–84* *Revise for C3 page 50 Worked exam style question 2(b)*
4 **(a)** Find the values of $\tan\theta$ which satisfy the equation $6\sec^2\theta - 3\tan^2\theta = 11\tan\theta$. **(b)** Hence solve, for $-180° < \theta < 180°$, $6\sec^2\theta - 3\tan^2\theta = 11\tan\theta$, giving your answers to the nearest 0.1°.	*Review Heinemann Book C3 pages 85–86*

5 Given that $\sin x \sin y - 2 \cos x \cos y = 3 \sin x \cos y$, and that $\cot x = k$, $k \neq -\frac{2}{3}$, express $\cot y$ in terms of k.

Review Heinemann Book C3 page 73
Revise for C3 page 50
Worked exam style question 2a

---

6 Given that $-\dfrac{\pi}{2} \leqslant \arcsin x \leqslant \dfrac{\pi}{2}$, find the exact value of $\cos\{\arcsin(-\frac{1}{3})\}$.

Review Heinemann Book C3 pages 87–90
Revise for C3 page 49
Example 3

---

7 **(a)** Sketch, for $-90° \leqslant x < 180°$, the curve with equation $y = \operatorname{cosec} x$.

(b) On the same set of axes and for the same interval sketch the curve with equation $y = \sec 2x$.

(c) Show that the x-coordinates of the points of the intersection of the curves satisfy the equation

$$2 \sin^2 x + \sin x - 1 = 0.$$

> You need to use an appropriate double angle formula (see Chapter 7).

Review Heinemann Book C3 pages 76–78 and 101
Revise for C3 page 50
Worked exam style question 3

Test yourself answers

1 (a) $\operatorname{cosec} \theta$ **(b)** $\tan x$ **2** 105°, 165°, 285°, 345° **3** $-\dfrac{\sqrt{13}}{2}$

4 (a) $\frac{2}{3}$, 3 **(b)** −146.3°, −108.4°, 33.7°, 71.6°

5 $\dfrac{1}{3 + 2k}$ **6** $\dfrac{2\sqrt{2}}{3}$

7 (a) (b)

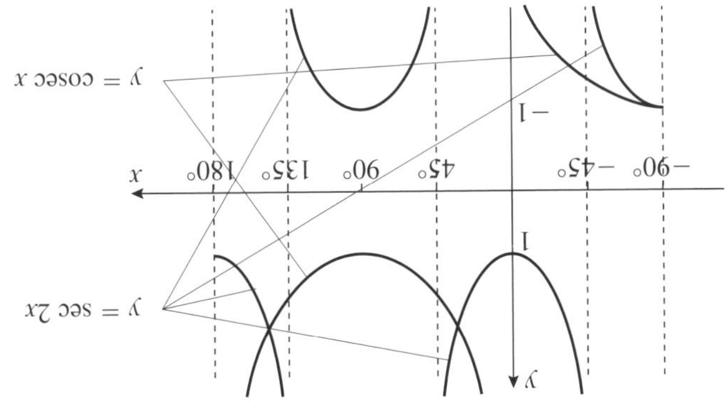

(c) Solving $\operatorname{cosec} x = \sec 2x$ will give the x-coordinates of points of intersection of the curves,
so $\cos 2x = \sin x$
$(1 - 2\sin^2 x) = \sin x$
$2\sin^2 x + \sin x - 1 = 0$

Further trigonometric identities and their application

7

Key points to remember

1 The addition (or compound angle) formulae are:

(i) $\sin(A + B) \equiv \sin A \cos B + \cos A \sin B$

(ii) $\sin(A - B) \equiv \sin A \cos B - \cos A \sin B$

(iii) $\cos(A + B) \equiv \cos A \cos B - \sin A \sin B$

(iv) $\cos(A - B) \equiv \cos A \cos B + \sin A \sin B$

(v) $\tan(A + B) \equiv \dfrac{\tan A + \tan B}{1 - \tan A \tan B}$

(vi) $\tan(A - B) \equiv \dfrac{\tan A - \tan B}{1 + \tan A \tan B}$

2 The double angle formulae are:

(i) $\sin 2A \equiv 2 \sin A \cos A$

(ii) $\cos 2A \equiv \cos^2 A - \sin^2 A \equiv 2\cos^2 A - 1 \equiv 1 - 2\sin^2 A$

(iii) $\tan 2A \equiv \dfrac{2 \tan A}{1 - \tan^2 A}$

3 (i) Expressions of the form $a \sin \theta \pm b \cos \theta$ can be rewritten in terms of a sine only or a cosine only (by using 'the R formulae').

[You are not expected to know the following results, but you need to be able to derive them for given values of a and b.

For positive values of a and b:

$a \sin \theta \pm b \cos \theta \equiv R \sin(\theta \pm \alpha)$, with $R > 0$ and $0 < \alpha < 90°$,
$a \cos \theta \pm b \sin \theta \equiv R \cos(\theta \mp \alpha)$, with $R > 0$ and $0 < \alpha < 90°$,
where $R \cos \alpha = a$, $R \sin \alpha = b$ and $R = \sqrt{a^2 + b^2}$.]

(ii) Equations of the form $a \cos \theta + b \sin \theta = c$, where a, b and c are constants, can always be solved by using 'the R formula'. Remember that if $c = 0$, the equation reduces to the form $\tan \theta = k$.

4 Products of sines and/or cosines can be expressed as the sum or difference of sines or cosines.

The following formulae are derived from the relevant addition formulae in **1**

$$2\sin A\cos B \equiv \sin(A+B)+\sin(A-B)$$
$$2\cos A\sin B \equiv \sin(A+B)-\sin(A-B)$$
$$2\cos A\cos B \equiv \cos(A+B)+\cos(A-B)$$
$$2\sin A\sin B \equiv -\{\cos(A+B)-\cos(A-B)\}$$

5 Sums or differences of sines or cosines can be expressed as a product of sines and/or cosines, using 'the factor formulae'. The following formulae are derived from **4**.

$$\sin P+\sin Q \equiv 2\sin\left(\frac{P+Q}{2}\right)\cos\left(\frac{P-Q}{2}\right)$$

$$\sin P-\sin Q \equiv 2\cos\left(\frac{P+Q}{2}\right)\sin\left(\frac{P-Q}{2}\right)$$

$$\cos P+\cos Q \equiv 2\cos\left(\frac{P+Q}{2}\right)\cos\left(\frac{P-Q}{2}\right)$$

$$\cos P-\cos Q \equiv -2\sin\left(\frac{P+Q}{2}\right)\sin\left(\frac{P-Q}{2}\right)$$

Example 1

Express as a single trigonometric function:

(a) $\sin 3x\cos x-\cos 3x\sin x$

(b) $\sqrt{1+\cos 4\theta}$

(c) $4\sin x\cos x\cos 2x$.

(a) $\sin 3x\cos x-\cos 3x\sin x = \sin(3x-x)$

$= \sin 2x$

Using **1** (ii)

(b) $\sqrt{1+\cos 4\theta}=\sqrt{1+(2\cos^2 2\theta-1)}$

$=\sqrt{2\cos^2 2\theta}$

$=\sqrt{2}\cos 2\theta$

Using **2** (ii): with $A = 2\theta$ [you need to choose the appropriate form of $\cos 2A$]

(c) $4\sin x\cos x\cos 2x = 2(2\sin x\cos x)\cos 2x$

$= 2\sin 2x\cos 2x$

$= \sin 4x$

Using **2** (i): with $A = x$

Using **2** (i): again with $A = 2x$

Example 2

Find, in surd form, the values of:

(a) $\cos 80° \cos 50° + \sin 80° \sin 50°$

(b) $(\cos 22\tfrac{1}{2}° - \sin 22\tfrac{1}{2}°)^2$.

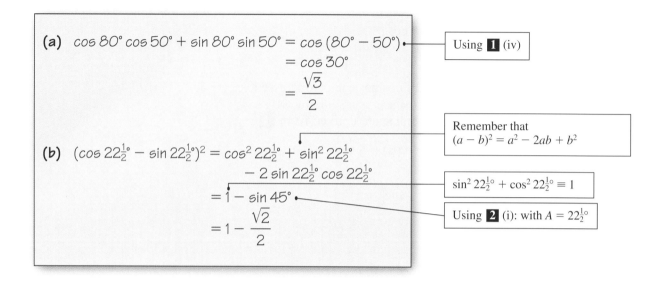

(a) $\cos 80° \cos 50° + \sin 80° \sin 50° = \cos(80° - 50°)$

$= \cos 30°$

$= \dfrac{\sqrt{3}}{2}$

Using **1** (iv)

(b) $(\cos 22\tfrac{1}{2}° - \sin 22\tfrac{1}{2}°)^2 = \cos^2 22\tfrac{1}{2}° + \sin^2 22\tfrac{1}{2}° - 2 \sin 22\tfrac{1}{2}° \cos 22\tfrac{1}{2}°$

$= 1 - \sin 45°$

$= 1 - \dfrac{\sqrt{2}}{2}$

Remember that $(a - b)^2 = a^2 - 2ab + b^2$

$\sin^2 22\tfrac{1}{2}° + \cos^2 22\tfrac{1}{2}° \equiv 1$

Using **2** (i): with $A = 22\tfrac{1}{2}°$

Example 3

Given that $\sin A = \frac{4}{5}$ and that A is obtuse, and $\cos B = -\frac{7}{25}$ and that B is reflex, find, as a fraction, the value of $\tan(A - B)$.

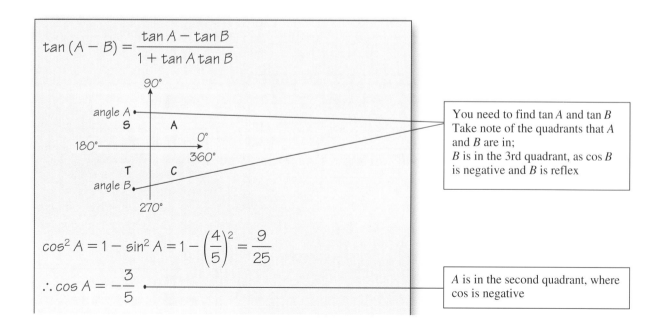

$$\tan(A - B) = \frac{\tan A - \tan B}{1 + \tan A \tan B}$$

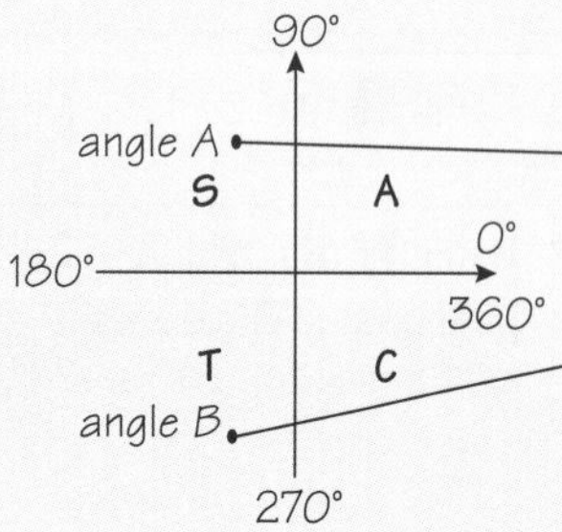

You need to find $\tan A$ and $\tan B$. Take note of the quadrants that A and B are in; B is in the 3rd quadrant, as $\cos B$ is negative and B is reflex

$$\cos^2 A = 1 - \sin^2 A = 1 - \left(\frac{4}{5}\right)^2 = \frac{9}{25}$$

$$\therefore \cos A = -\frac{3}{5}$$

A is in the second quadrant, where cos is negative

so $\tan A = \dfrac{\frac{4}{5}}{-\frac{3}{5}} = -\dfrac{4}{3}$

Using $\tan A = \dfrac{\sin A}{\cos A}$

As $\cos B = -\dfrac{7}{25}$

so $\sec B = -\dfrac{25}{7}$

and $\tan^2 B = \left(-\dfrac{25}{7}\right)^2 - 1 = \dfrac{576}{49}$

Using $1 + \tan^2 B \equiv \sec^2 B$

so $\tan B = \dfrac{24}{7}$

B is in the third quadrant, where tan is positive

so $\tan (A - B) = \dfrac{(-\frac{4}{3}) - \frac{24}{7}}{1 + (-\frac{4}{3})(\frac{24}{7})}$

$= \left(-\dfrac{100}{21}\right) \div \left(-\dfrac{75}{21}\right)$

$= \dfrac{4}{3}$

You can also work with the associated acute angles A' and B'. Draw right angled triangles and use Pythagoras' theorem.

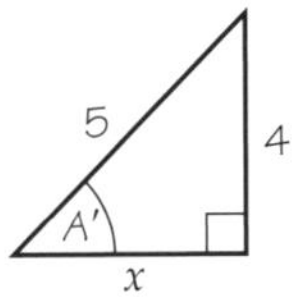

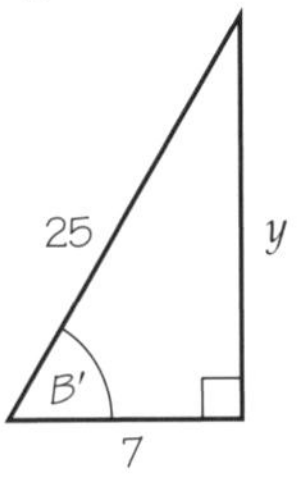

$x^2 = 5^2 - 4^2$
so $x = 3$
As A is in 2nd quad,
$\sin A = \sin A' = \frac{4}{5}$
$\cos A = -\cos A' = -\frac{3}{5}$
$\tan A = -\tan A' = -\frac{4}{3}$

$y^2 = 25^2 - 7^2$
so $y = 24$
As B is in 3rd quad,
$\sin B = -\sin B' = -\frac{24}{25}$
$\cos B = -\cos B' = -\frac{7}{25}$
$\tan B = \tan B' = \frac{24}{7}$

Example 4

Given that $2\sin(x - y) = 5\cos(x + y)$, and that $\tan x = -3$, find $\tan y$.

Expanding

$$2\sin x\cos y - 2\cos x\sin y = 5\cos x\cos y - 5\sin x\sin y$$

so $$\frac{2\sin x\cos y}{\cos x\cos y} - \frac{2\cos x\sin y}{\cos x\cos y} = \frac{5\cos x\cos y}{\cos x\cos y} - \frac{5\sin x\sin y}{\cos x\cos y}$$

$$2\tan x - 2\tan y = 5 - 5\tan x\tan y$$

so $$-6 - 2\tan y = 5 + 15\tan y$$

so $$17\tan y = -11$$

$$\tan y = -\frac{11}{17}$$

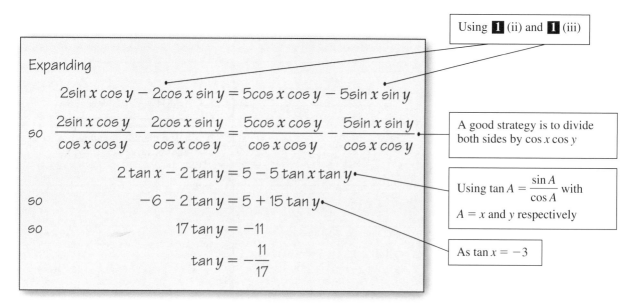

Using **1** (ii) and **1** (iii)

A good strategy is to divide both sides by $\cos x\cos y$

Using $\tan A = \dfrac{\sin A}{\cos A}$ with $A = x$ and y respectively

As $\tan x = -3$

Example 5

Solve, in the interval $0 \leqslant \theta \leqslant 360°$, $2\cos 2\theta + \sin\theta + 1 = 0$.

Using $\cos 2\theta \equiv 1 - 2\sin^2\theta$, the given equation becomes

$$2(1 - 2\sin^2\theta) + \sin\theta + 1 = 0.$$

or $$4\sin^2\theta - \sin\theta - 3 = 0$$

$$(4\sin\theta + 3)(\sin\theta - 1) = 0$$

so $\sin\theta = -\frac{3}{4}$ or $\sin\theta = 1$

Solutions are: 90°, 228.6°, 311.4° (1 d.p.)

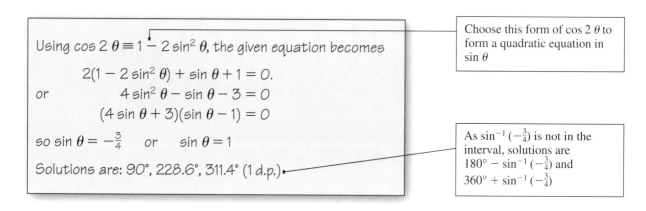

Choose this form of $\cos 2\theta$ to form a quadratic equation in $\sin\theta$

As $\sin^{-1}(-\frac{3}{4})$ is not in the interval, solutions are $180° - \sin^{-1}(-\frac{3}{4})$ and $360° + \sin^{-1}(-\frac{3}{4})$

Worked exam style question 1

(a) Express $\sqrt{3}\sin\theta - \cos\theta$ in the form $R\sin(\theta - \alpha)$, $R > 0$ and $0 \leqslant \alpha < \dfrac{\pi}{2}$.

(b) Hence sketch, in the interval $-\pi \leqslant \theta \leqslant 2\pi$, the graph of $y = \sqrt{3}\sin\theta - \cos\theta$, showing the coordinates of points of intersection with the axes.

(c) Solve, in the interval $-\pi \leqslant \theta \leqslant 2\pi$, $\sqrt{3}\sin\theta - \cos\theta = -1$, giving your answers in terms of π.

(a) Setting $\sqrt{3}\sin\theta - \cos\theta \equiv R\sin(\theta - \alpha)$,

so $\sqrt{3}\sin\theta - \cos\theta \equiv R\sin\theta\cos\alpha - R\cos\theta\sin\alpha$

Using **1** (ii): check with **3**

$\Rightarrow R\cos\alpha = \sqrt{3}$ and $R\sin\alpha = 1$

Comparing coefficients of $\sin\theta$ and $\cos\theta$ respectively

$\Rightarrow R = 2,\ \alpha = \dfrac{\pi}{6}$

$R = \sqrt{(\sqrt{3})^2 + 1^2}$, $\tan\alpha = \dfrac{1}{\sqrt{3}}$

so $\sqrt{3}\sin\theta - \cos\theta \equiv 2\sin\left(\theta - \dfrac{\pi}{6}\right)$

(b)

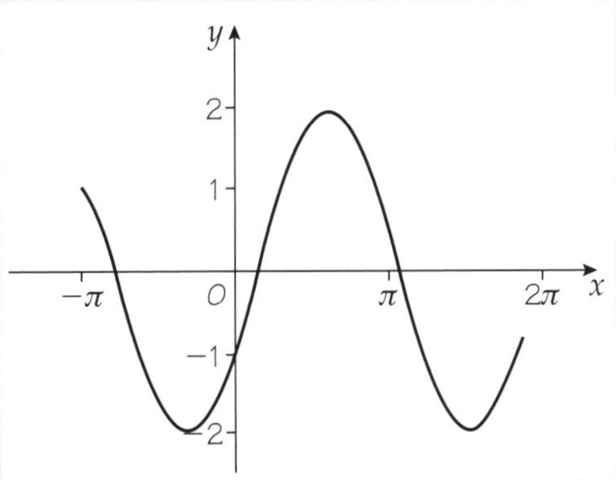

The curve is a horizontal translation of $y = \sin\theta$ by $\left(\dfrac{\pi}{6}\right)$ to the right, and a vertical stretch with scale factor 2

(c) Solving $2\sin\left(\theta - \dfrac{\pi}{6}\right) = -1$ in $-\dfrac{7\pi}{6} \leqslant \left(\theta - \dfrac{\pi}{6}\right) \leqslant \dfrac{11\pi}{6}$

so $\sin\left(\theta - \dfrac{\pi}{6}\right) = -\frac{1}{2}$

$\Rightarrow \left(\theta - \dfrac{\pi}{6}\right) = -\dfrac{5\pi}{6}, -\dfrac{\pi}{6}, \dfrac{7\pi}{6}, \dfrac{11\pi}{6}$

Solutions are in the 3rd and 4th quadrants as sine is negative

so $\theta = -\dfrac{2\pi}{3}, 0, \dfrac{4\pi}{3}, 2\pi$

The solutions are the θ-coordinates of the points of intersection of the line $y = -1$ with $y = 3\sin\theta - \cos\theta$ [look back to the graph]

Worked exam style question 2

(a) Using the formula for $\tan(A + B)$ show that

$$\tan 2\theta \equiv \frac{2\tan\theta}{1 - \tan^2\theta}.$$

(b) By choosing a suitable value of θ show that $\tan\left(\frac{3\pi}{8}\right) = \sqrt{2} + 1$.

(a) Using $\tan(A + B) \equiv \dfrac{\tan A + \tan B}{1 - \tan A \tan B}$ with $A = B = \theta$

gives $\tan(\theta + \theta) \equiv \dfrac{\tan\theta + \tan\theta}{1 - \tan\theta\tan\theta}$

so $\tan 2\theta \equiv \dfrac{2\tan\theta}{1 - \tan^2\theta}$

(b) Let $\theta = \dfrac{3\pi}{8}$

so $\tan\dfrac{3\pi}{4} = \dfrac{2\tan\alpha}{1 - \tan^2\alpha}$, where $\alpha = \dfrac{3\pi}{8}$

$\Rightarrow -1 = \dfrac{2\tan\alpha}{1 - \tan^2\alpha}$

so $\tan^2\alpha - 2\tan\alpha - 1 = 0$

$\tan\alpha = \dfrac{2 \pm \sqrt{8}}{2}$ — Using the quadratic formula

but $\tan\alpha = \tan\dfrac{3\pi}{8}$ which is positive — $\dfrac{3\pi}{8}$ is in 1st quadrant, so tan is positive

so $\tan\left(\dfrac{3\pi}{8}\right) = \dfrac{2 + \sqrt{8}}{2} = \dfrac{2 + 2\sqrt{2}}{2} = \sqrt{2} + 1$ — $\sqrt{8} = \sqrt{4 \times 2} = \sqrt{4} \times \sqrt{2} = 2\sqrt{2}$

Worked exam style question 3

Prove that $\cot\theta + \operatorname{cosec}\theta \equiv \cot\left(\frac{\theta}{2}\right)$, provided $\theta \neq \frac{n\pi}{4}$, $n \in Z$.

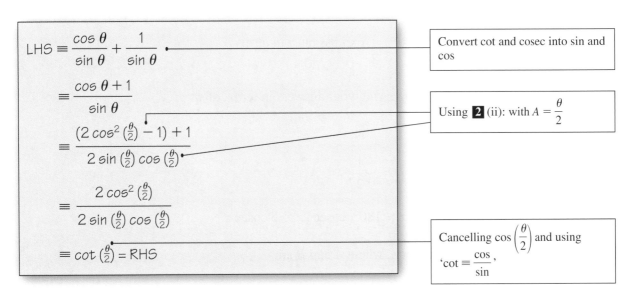

$$\text{LHS} \equiv \frac{\cos\theta}{\sin\theta} + \frac{1}{\sin\theta}$$

Convert cot and cosec into sin and cos

$$\equiv \frac{\cos\theta + 1}{\sin\theta}$$

$$\equiv \frac{(2\cos^2\left(\frac{\theta}{2}\right) - 1) + 1}{2\sin\left(\frac{\theta}{2}\right)\cos\left(\frac{\theta}{2}\right)}$$

Using **2** (ii): with $A = \frac{\theta}{2}$

$$\equiv \frac{2\cos^2\left(\frac{\theta}{2}\right)}{2\sin\left(\frac{\theta}{2}\right)\cos\left(\frac{\theta}{2}\right)}$$

$$\equiv \cot\left(\frac{\theta}{2}\right) = \text{RHS}$$

Cancelling $\cos\left(\frac{\theta}{2}\right)$ and using '$\cot \equiv \frac{\cos}{\sin}$'

Revision exercise 7

1 State which of the following are equivalent to $\cos\theta$.

(a) $\cos^2\left(\frac{\theta}{2}\right) - \sin^2\left(\frac{\theta}{2}\right)$

(b) $2\sin\left(\frac{\theta}{2}\right)\cos\left(\frac{\theta}{2}\right)$

(c) $2\sin^2\left(\frac{\theta}{2}\right) - 1$

(d) $\cos 2\theta\cos\theta + \sin 2\theta\sin\theta$

2 (a) Express the following as a single trigonometric ratio.

(i) $\sin 10° \cos 50° + \cos 10° \sin 50°$

(ii) $2\cos^2 75° - 1$

(iii) $\dfrac{\tan 70° - \tan 25°}{1 + \tan 70° \tan 25°}$

(b) Hence write down the exact value of each part of **(a)**.

3 Prove that:

(a) $\cos^2 2\theta \equiv \dfrac{1 + \cos 4\theta}{2}$

(b) $(\cos\theta + \sin\theta)^2 - (\cos\theta - \sin\theta)^2 = 2\sin 2\theta$.

4 Given that $x = 1 + \cos\theta$ and $y = \cos 2\theta$, show that $y = 2x^2 - 4x + 1$.

5 Given that $\tan(x - 45°) = k$, $k \neq 1$, express $\tan x$ in terms of k.

6 Given that $\sin(x + y) = 3\cos(x - y)$, express $\tan y$ in terms of $\tan x$.

7 Solve, in the interval $0 \leqslant x \leqslant \pi$, giving your answers in terms of π:

(a) $1 + 2\sin x \cos x = 0$

(b) $\tan 2x = 3\tan x$

(c) $\cos x \cos\left(\frac{\pi}{5}\right) + \sin x \sin\left(\frac{\pi}{5}\right) = -\frac{1}{2}$.

8 Solve, in the interval $-180° \leqslant x \leqslant 180°$, $\operatorname{cosec} 2x = 3\operatorname{cosec} x$.

9 Given that $\cos A = \frac{3}{5}$ and $\sin B = \frac{1}{3}$, where A and B are both acute, find the exact values of:

(a) $\sin A$ **(b)** $\operatorname{cosec} 2A$

(c) $\cos 2B$ **(d)** $\sin(A - B)$.

10 Given that $2\cos(\theta + 60°) \equiv k\cos\theta - c\sin(\theta + 60°)$, where c and k are constants, find the values of c and k.

11 **(a)** By expanding $\sin(45° - 30°)$, show that

$$\sin 15° = \frac{\sqrt{2}(\sqrt{3} - 1)}{4}.$$

(b) In the triangle shown, show that $AB = 3(\sqrt{3} - 1)$ cm.

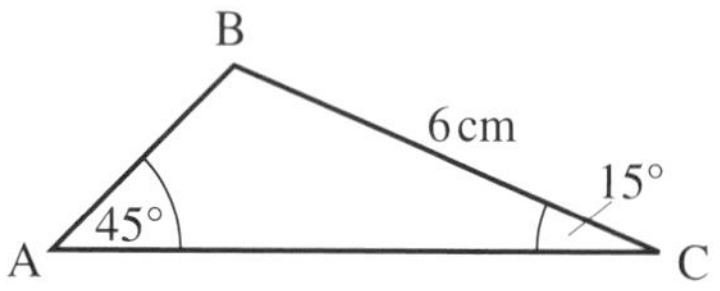

12 **(a)** Given that $\cos 2\theta = \frac{5}{6}$, find possible values of $\sin\theta$.

(b) Given that $\tan 2\theta = -\frac{3}{4}$, and that θ is obtuse, find the value of **(i)** $\tan\theta$ **(ii)** $\sec\theta$.

13 **(a)** Given that $\tan A = -\frac{4}{3}$, where A is obtuse, show that $\sin 2A = -\frac{24}{25}$.

(b) Given also that $\tan B = \frac{5}{12}$, where B is reflex, find the exact values of **(i)** $\tan(A - B)$ **(ii)** $\sin(A + B)$.

14 Prove the following identities.

(a) $\cot 2\theta - \tan 2\theta \equiv 2\cot 4\theta$

(b) $1 + 2\cos^2\theta - 4\sin^2\theta \equiv 3\cos 2\theta$

(c) $\tan A + \cot B \equiv \sec A \operatorname{cosec} B \cos(A - B)$

15 Solve, in the interval $0 \leqslant x < 360°$, the following equations, giving your answers to the nearest 0.1°.

(a) $\cos 2x = 2 \cos x$

(b) $\sin x = 2 \cos (x - 60°)$

(c) $\sin x \cos x = 1 - \cos 2x$

16 **(a)** Show that $\sin x + \cos x \equiv \sqrt{2} \sin (x + 45°)$.

(b) Sketch the curve with equation $y = \sin x + \cos x$, $0 \leqslant x \leqslant 360°$, showing the coordinates of any maximum or minimum points.

17 **(a)** Express $5 \cos x + 4 \sin x$ in the form $R \cos (x - \alpha)$, where $R > 0$ and $0 < \alpha < 90°$. Give your value of α to 3 significant figures.

(b) Calculate, to the nearest 0.1°, the solutions of $5 \cos x + 4 \sin x = 2$, in the interval $0 \leqslant x \leqslant 360°$.

18 Solve, in the interval $-180° < x < 180°$, the following equations. Give your answers to the nearest 0.1°.

(a) $2 \sin x + 3 \cos x = 0$

(b) $2 \sin x + 3 \cos x = 1.5$

(c) $2 \sin x + 3 \cos 2x = 0$

19 **(a)** Using the formula for $\cos (A + B)$ and $\cos (A - B)$ show that

$$\cos X + \cos Y \equiv 2 \cos \left(\frac{X + Y}{2}\right) \cos \left(\frac{X - Y}{2}\right) \text{ and}$$

$$\cos X - \cos Y \equiv -2 \sin \left(\frac{X + Y}{2}\right) \sin \left(\frac{X - Y}{2}\right).$$

(b) Prove that $\dfrac{\cos 3x + \cos x}{\cos x - \cos 3x} \equiv \cot 2x \cot x$.

(c) Solve, in the interval $0 \leqslant x \leqslant 180°$, $\cos 5x = -\cos x$.

20 **(a)** Using the formula for $\sin (A + B)$ and $\sin (A - B)$ show that

$$\sin X + \sin Y \equiv 2 \sin \left(\frac{X + Y}{2}\right) \cos \left(\frac{X - Y}{2}\right).$$

(b) Prove that $\dfrac{\sin 15° + \sin 45°}{\cos 15° + \cos 45°} = \dfrac{\sqrt{3}}{3}$.

Test yourself

What to review

If your answer is incorrect

1 Simplify the following expressions.

(a) $\sqrt{1 - \cos 2\theta}$

(b) $\sin 6\theta \cos \theta - \cos 6\theta \sin \theta$

Review Heinemann Book C3 pages 95 and 101
Revise for C3 page 57 Example 1

---

2 (a) Given that $\sin A = \frac{5}{13}$, and that A is an acute angle, find the exact values of:
(i) $\sin 2A$ **(ii)** $\tan 2A$.

(b) Given also that $\cos B = \frac{7}{25}$, and that B is a reflex angle, show that $\sin(A + B) = -\frac{253}{325}$.

Review Heinemann Book C3 page 103
Revise for C3 page 58 Example 3

---

3 (a) Prove that $\tan \theta + \cot \theta = 2 \operatorname{cosec} 2\theta$.

(b) Deduce the exact value of $\tan\left(\frac{\pi}{8}\right) + \cot\left(\frac{\pi}{8}\right)$.

Review Heinemann Book C3 page 105
Revise for C3 page 62 Worked exam style question 2

---

4 Solve, in the interval $0 \leqslant x \leqslant 360°$,

$\cos 2x - 3\cos x + 2 = 0$.

Review Heinemann Book C3 page 106
Revise for C3 page 60 Example 5

---

5 (a) Show that $\cos \theta - 2\sin \theta$ can be written in the form $R\cos(\theta + \alpha)$, where $R > 0$, and $0 < \alpha < \frac{\pi}{2}$.
Give R in surd form and α in radians to 4 significant figures.

(b) Solve, in the interval $-\pi < \theta < \pi$, the equation $\cos \theta - 2\sin \theta = -0.6$, giving answers to 3 significant figures.

Review Heinemann Book C3 page 110
Revise for C3 page 61 Worked exam style question 1

---

6 Prove that $1 + \sin \theta - \cos \theta \equiv 2\sin(\frac{1}{2}\theta)[\cos(\frac{1}{2}\theta) + \sin(\frac{1}{2}\theta)]$.

Review Heinemann Book C3 pages 101–102
Revise for C3 page 63 Worked exam style question 3

Test yourself answers

1 (a) $\sqrt{2}\sin\theta$ **(b)** $\sin 5\theta$ **2 (a) (i)** $\frac{120}{169}$ **(ii)** $\frac{120}{119}$ **3 (b)** $2\sqrt{2}$ **4** 0°, 60°, 300°, 360°

5 (a) $R = \sqrt{5}$, $\alpha = 1.107$ **(b)** -2.95, 0.735

6 $\sin\theta + 1 - \cos\theta = 2\sin(\frac{1}{2}\theta)\cos(\frac{1}{2}\theta) + 2\sin^2(\frac{1}{2}\theta)$
$= 2\sin(\frac{1}{2}\theta)[\cos(\frac{1}{2}\theta) + \sin(\frac{1}{2}\theta)]$

Differentiation 8

Key points to remember

1 You can use the chain rule to differentiate a function of a function.

If $y = [f(x)]^n$ then $\frac{dy}{dx} = n[f(x)]^{n-1}f'(x)$.

If $y = f[g(x)]$ then $\frac{dy}{dx} = f'[g(x)]g'(x)$.

2 Another form of the **chain rule** states that $\frac{dy}{dx} = \frac{dy}{du} \times \frac{du}{dx}$ where y is a function of u, and u is a function of x.

3 A particular case of the chain rule is the result $\frac{dy}{dx} = \frac{1}{\left(\frac{dx}{dy}\right)}$.

4 You can use the product rule when two functions $u(x)$ and $v(x)$ are multiplied together.

If $y = uv$ then $\frac{dy}{dx} = u\frac{dv}{dx} + v\frac{du}{dx}$.

5 You can use the quotient rule when one function $u(x)$ is divided by another function $v(x)$, to form a rational function.

If $y = \frac{u}{v}$ then $\frac{dy}{dx} = \frac{v\frac{du}{dx} - u\frac{dv}{dx}}{v^2}$.

6 If $y = e^x$, then $\frac{dy}{dx} = e^x$ also and if $y = e^{f(x)}$, then $\frac{dy}{dx} = f'(x)e^{f(x)}$.

7 If $y = \ln x$ then $\frac{dy}{dx} = \frac{1}{x}$ and if $y = \ln[f(x)]$, then $\frac{dy}{dx} = \frac{f'(x)}{f(x)}$.

8 If $y = \sin x$ then $\frac{dy}{dx} = \cos x$ and if $y = \sin f(x)$ then $\frac{dy}{dx} = f'(x)\cos f(x)$.

9 If $y = \cos x$ then $\frac{dy}{dx} = -\sin x$

and if $y = \cos f(x)$ then $\frac{dy}{dx} = -f'(x) \sin f(x)$.

10 If $y = \tan x$ then $\frac{dy}{dx} = -\sec^2 x$

and if $y = \tan f(x)$ then $\frac{dy}{dx} = f'(x) \sec^2 f(x)$.

11 If $y = \operatorname{cosec} x$ then $\frac{dy}{dx} = -\operatorname{cosec} x \cot x$

and if $y = \operatorname{cosec} f(x)$ then $\frac{dy}{dx} = -f'(x) \operatorname{cosec} f(x) \cot f(x)$.

12 If $y = \sec x$ then $\frac{dy}{dx} = \sec x \tan x$

and if $y = \sec f(x)$ then $\frac{dy}{dx} = f'(x) \sec f(x) \tan f(x)$.

13 If $y = \cot x$ then $\frac{dy}{dx} = -\operatorname{cosec}^2 x$

and if $y = \cot f(x)$ then $\frac{dy}{dx} = -f'(x) \operatorname{cosec}^2 f(x)$.

Example 1

Given that $y = (3x^2 + 5)^{\frac{1}{3}}$ find $\frac{dy}{dx}$, using the chain rule.

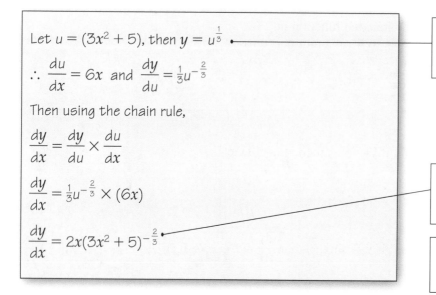

Let $u = (3x^2 + 5)$, then $y = u^{\frac{1}{3}}$

$\therefore \frac{du}{dx} = 6x$ and $\frac{dy}{du} = \frac{1}{3}u^{-\frac{2}{3}}$

Then using the chain rule,

$$\frac{dy}{dx} = \frac{dy}{du} \times \frac{du}{dx}$$

$$\frac{dy}{dx} = \frac{1}{3}u^{-\frac{2}{3}} \times (6x)$$

$$\frac{dy}{dx} = 2x(3x^2 + 5)^{-\frac{2}{3}}$$

Using **2**: put the bracket equal to u and use the chain rule in terms of u

Ensure that you give your answer in terms of x, with no u terms present

You could have used the chain rule stated in **1** to differentiate without a substitution

Example 2

Find the value of $\frac{dy}{dx}$ at the point (3, 1) on the curve with equation

$5y^3 - \frac{2}{y} = x$.

As $x = 5y^3 - 2y^{-1}$

$$\frac{dx}{dy} = 15y^2 + \frac{2}{y^2}$$

$$= \frac{15y^4 + 2}{y^2}$$

$$\therefore \frac{dy}{dx} = \frac{y^2}{15y^4 + 2}$$

$$= \frac{1}{17}$$

Using **3**: start with $x = 5y^3 - 2y^{-1}$ and differentiate with respect to y

Use $\frac{dy}{dx} = \frac{1}{\frac{dx}{dy}}$ and substitute $y = 1$

Example 3

Given that $f(x) = x^5\sqrt{(x^2 + 4)}$, find $f'(x)$.

Using **4**: recognise that this is a product of two functions

Let $u = x^5$ and $v = \sqrt{(x^2 + 4)} = (x^2 + 4)^{\frac{1}{2}}$

Then $\frac{du}{dx} = 5x^4$ and $\frac{dv}{dx} = 2x \times \frac{1}{2}(x^2 + 4)^{-\frac{1}{2}}$

The second function is a function of a function that requires the chain rule

Using $\frac{dy}{dx} = u\frac{dv}{dx} + v\frac{du}{dx}$

$$= x^5 \times x(x^2 + 4)^{-\frac{1}{2}} + \sqrt{(x^2 + 4)} \times 5x^4$$

$$= \frac{x^6 + 5x^4(x^2 + 4)}{\sqrt{(x^2 + 4)}}$$

$$= \frac{6x^6 + 20x^4}{\sqrt{(x^2 + 4)}}$$

$$= \frac{2x^4(3x^2 + 10)}{\sqrt{(x^2 + 4)}}$$

Collect terms to simplify, and factorise to give the final answer

Example 4

Given that $y = \dfrac{3x}{x^2+1}$ find $\dfrac{dy}{dx}$.

Let $u = 3x$ and $v = x^2 + 1$

$\dfrac{du}{dx} = 3$ and $\dfrac{dv}{dx} = 2x$

Using $\dfrac{dy}{dx} = \dfrac{v\dfrac{du}{dx} - u\dfrac{dv}{dx}}{v^2}$

Using **5**: recognise that this is a quotient and use the quotient rule

$= \dfrac{(x^2+1).3 - 3x.2x}{(x^2+1)^2}$

$= \dfrac{3 - 3x^2}{(x^2+1)^2}$

Simplify the numerator of the fraction

$= \dfrac{3(1-x)(1+x)}{(x^2+1)^2}$

Factorise, using the difference of two squares

Example 5

Find the function $f'(x)$ where $f(x)$ is:

(a) $\sin 2x + \cos 3x$

(b) $4 \tan^2 5x$

(c) $e^{3x} \sec 7x$

(d) $\ln(\operatorname{cosec} x + \cot x)$.

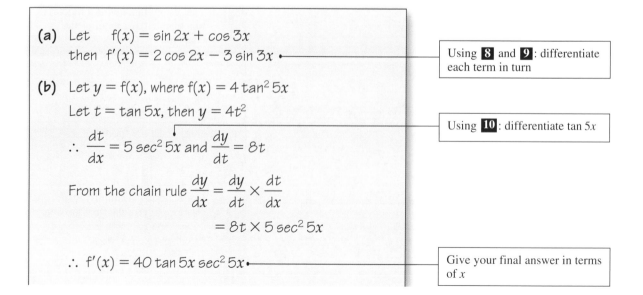

(a) Let $f(x) = \sin 2x + \cos 3x$
then $f'(x) = 2\cos 2x - 3\sin 3x$

Using **8** and **9**: differentiate each term in turn

(b) Let $y = f(x)$, where $f(x) = 4\tan^2 5x$

Let $t = \tan 5x$, then $y = 4t^2$

$\therefore \dfrac{dt}{dx} = 5\sec^2 5x$ and $\dfrac{dy}{dt} = 8t$

Using **10**: differentiate $\tan 5x$

From the chain rule $\dfrac{dy}{dx} = \dfrac{dy}{dt} \times \dfrac{dt}{dx}$

$= 8t \times 5\sec^2 5x$

$\therefore f'(x) = 40\tan 5x \sec^2 5x$

Give your final answer in terms of x

(c) Let $f(x) = e^{3x} \sec 7x$

Use the product rule to give

$$f'(x) = e^{3x}\, 7 \sec 7x \tan 7x + \sec 7x \times 3e^{3x}$$
$$= e^{3x} \sec 7x\,(7 \tan 7x + 3)$$

Using **12**: differentiate sec $7x$

(d) Let $f(x) = \ln(\operatorname{cosec} x + \cot x)$

then $$f'(x) = (-\operatorname{cosec} x \cot x - \operatorname{cosec}^2 x) \times \frac{1}{(\operatorname{cosec} x + \cot x)}$$
$$= -\operatorname{cosec} x\,(\operatorname{cosec} x + \cot x) \times \frac{1}{(\operatorname{cosec} x + \cot x)}$$
$$= -\operatorname{cosec} x$$

Using **11** and **13**: differentiate $\operatorname{cosec} x + \cot x$

Factorise and simplify the fraction

Worked exam style question 1

$f(x) = \sin^2 x - 3\cos^2 x,\ 0 < x < \frac{\pi}{2}$

(a) Find $f'(x)$.

The curve C with equation $y = f(x)$ crosses the x-axis at the point A.

(b) Find an equation for the tangent to C at A.

(a) $$f'(x) = 2 \sin x \cos x + 6 \cos x \sin x,$$
$$= 8 \sin x \cos x$$
$$= 4 \sin 2x$$

Using **1** with **8** and **9**

(b) At A, $$\sin^2 x - 3\cos^2 x = 0$$
$$\tan^2 x = 3$$
$$\tan x = \sqrt{3}$$
$$x = \frac{\pi}{3}$$

Find the coordinates of point A, by putting $y = 0$

At $x = \frac{\pi}{3}$, gradient $= 2\sqrt{3}$

Substitute $x = \frac{\pi}{3}$ into $f'(x)$

Equation of tangent is $y - 0 = 2\sqrt{3}\left(x - \frac{\pi}{3}\right)$

Use $y - y_1 = m(x - x_1)$

Worked exam style question 2

Given that P is the size of a population, that t is the time in minutes during which the population has been growing, and that $P = 1000 \ln(2t + 1)$, $t \geqslant 0$.

(a) Find $\dfrac{dP}{dt}$.

(b) Find the value of t at which the population is increasing at a rate of 100 units per minute.

(a) $\dfrac{dP}{dt} = 1000 \times \dfrac{2}{2t+1} = \dfrac{2000}{2t+1}$

Using **7**

(b) Let $\dfrac{2000}{2t+1} = 100$

then $2t + 1 = 20$

so $t = 9.5$

Revision exercise 8

1 Given that T is a measure of temperature in degrees C, t is the time in minutes and that $T = 20 + 60e^{-0.1t}$, $t \geqslant 0$.

(a) Find $\dfrac{dT}{dt}$.

(b) Find the value of T at which the temperature is decreasing at a rate of 1.8 °C per minute. *E*

2

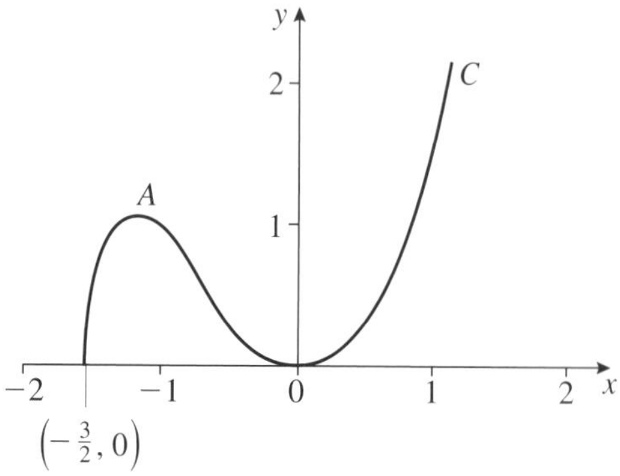

The curve C has equation $y = x^2(2x + 3)^{\frac{1}{2}}$, $x > -\frac{3}{2}$. Use calculus to find the x-coordinate of the maximum point A on the curve.

3 The curve C has equation $y = \dfrac{x}{4 + x^2}$. Use calculus to find the coordinates of the turning points of C. **(E)**

4 Given that $\mathrm{f}(x) = \dfrac{x^2 + 3x + 3}{x + 3}$, solve the equation $\mathrm{f}'(x) = \dfrac{22}{25}$. **(E)**

5 Given that $x = (y + 3)\mathrm{e}^y$:

(a) find $\dfrac{\mathrm{d}x}{\mathrm{d}y}$,

(b) use your answer to part **(a)** to find the value of $\dfrac{\mathrm{d}y}{\mathrm{d}x}$ at $y = 0$.

6 The curve C with equation $y = 2\mathrm{e}^x + 5$ meets the y-axis at the point M.

(a) Find the equation of the normal to C at M in the form $ax + by = c$, where a, b and c are integers.

(b) The normal to C at M crosses the x-axis at the point $N(n, 0)$, show that $n = 14$. **(E)**

7 $\mathrm{f}(x) = x + \dfrac{\mathrm{e}^x}{5}, x \in R$

(a) Find $\mathrm{f}'(x)$.

The curve C with equation $y = \mathrm{f}(x)$ crosses the y-axis at the point A.

(b) Find an equation for the tangent to C at A. **(E)**

8 The curve C has equation $y = \mathrm{f}(x)$, where

$$\mathrm{f}(x) = 3 \ln x + \frac{1}{x}, x > 0.$$

The point P is a stationary point on C.

(a) Calculate the x-coordinate of P.

(b) Show that the y-coordinate of P may be expressed in the form $k - k \ln k$, where k is a constant to be found. **(E)**

9 The function f is defined for real values of x by

$$\mathrm{f}(x) = \ln (x^2 + 1) - x.$$

Show, by differentiation, that $\mathrm{f}(x)$ is not an increasing function of x, and find the coordinates of the point of inflection.

10 Given that $y = \tan 2x - \cos^2 x$, find the exact value of $\dfrac{\mathrm{d}y}{\mathrm{d}x}$ at $x = \dfrac{\pi}{6}$.

11 Given that $x = \tan \frac{1}{2}y$, prove that $\frac{dy}{dx} = \frac{2}{1 + x^2}$. (E)

12 Use the derivatives of $\sec x$ and $\tan x$ to prove that
$\frac{d}{dx}[\ln(\sec x + \tan x)] = \sec x$.

13 Given that $y = e^{-2x}\cos 3x$ find $\frac{dy}{dx}$ and $\frac{d^2y}{dx^2}$ and write the second derivative in the form $Re^{-2x}\sin(3x - \alpha)$.

14

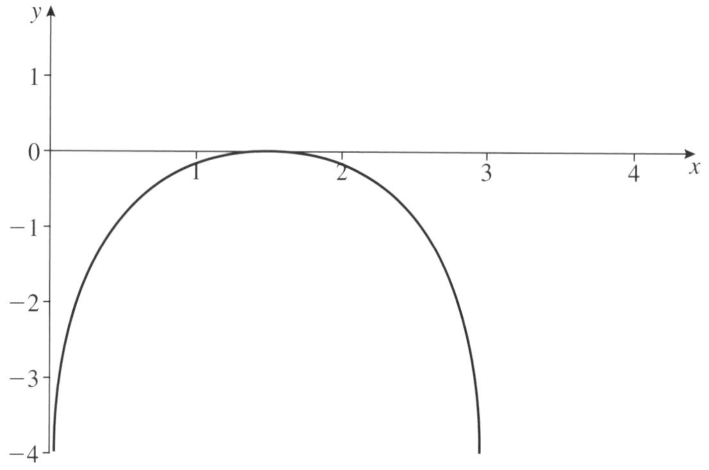

The curve shown above is part of the graph of $y = \ln \sin x$. Find the equation of the tangent to this curve at the point where $x = \frac{\pi}{4}$ and find where this tangent meets the x-axis.

Test yourself

1 Find $f'(x)$ for the following functions $f(x)$.

(a) $f(x) = (3x + 2)^6$

(b) $f(x) = (x^2 - 2x)^{-2}$

2 Find $f'(x)$ for the following functions of $f(x)$.

(a) $f(x) = x^3(3x + 2)^{\frac{1}{3}}$

(b) $f(x) = \frac{x + 1}{x^2 + 1}$

What to review

If your answer is incorrect

Review Heinemann Book C3 pages 120–121
Revise for C3 page 68 Example 1

Review Heinemann Book C3 pages 122–125
Revise for C3 pages 69–70 Examples 3 and 4

3 Find $\frac{dy}{dx}$, given that:

(a) $y = 5e^{3x-1}$

(b) $y = \ln(2x - 5)$

(c) $y = \ln\frac{x-5}{x+5}$.

Review Heinemann Book C3 pages 125–128
Revise for C3 page 70 Example 5

---

4 Differentiate with respect to x:

(a) $\sin 2x - 4\cos 3x$

(b) $3\sin^2 5x$

(c) $\tan x \sec x$.

Review Heinemann Book C3 pages 129–133
Revise for C3 page 70 Example 5

---

5 Find the value of $\frac{dy}{dx}$ at the point with coordinates (e, e) on the curve with equation $x = y\ln y$.

Review Heinemann Book C3 pages 121–122
Revise for C3 page 69 Example 2

Test yourself answers

1 **(a)** $f'(x) = 18(3x+2)^5$ **(b)** $f'(x) = -2(2x-2)(x^2-2x)^{-3}$

2 **(a)** $f'(x) = x^3(3x+2)^{-\frac{2}{3}} + 3x^2(3x+2)^{\frac{1}{3}} = 2x^2(3x+2)^{-\frac{2}{3}}[5x+3]$ **(b)** $f'(x) = \frac{1-2x-x^2}{(x^2+1)^2}$

3 **(a)** $\frac{dy}{dx} = 15e^{3x-1}$ **(b)** $\frac{dy}{dx} = \frac{2}{2x-5}$ **(c)** $\frac{dy}{dx} = \frac{1}{x-5} - \frac{1}{x+5} = \frac{10}{(x^2-25)}$

4 **(a)** $2\cos 2x + 12\sin 3x$ **(b)** $30\sin 5x\cos 5x$ or $15\sin 10x$ **(c)** $\sec x(\tan^2 x + \sec^2 x)$

5 $\frac{1}{2}$

Examination style paper

You must show sufficient working to make your methods clear.
Answers without working may gain no credit.

1 The amount, p units, of radioactivity present in a substance at time t seconds is given by the equation

$$p = 25\,e^{-0.006t}$$

(a) Write down the number of units present at time $t = 0$. **(1 mark)**

(b) Find the value of t when $p = 8$. **(3 marks)**

(c) Find the rate of decay of p, in units per second, at the instant that $t = 20$. **(4 marks)**

2 Solve, in the interval $0 < \theta < 360°$, the equation $\cot^2\theta = 1 - 4\,\text{cosec}\,\theta$, giving your answer to the nearest 0.1°. **(6 marks)**

3 The straight line with equation $x + y = 2$ crosses the curve with equation $y = 3\ln x$, $x > 0$, at the point P.

(a) Show that the x-coordinate of P satisfies the equation

$$x = e^{\left(\frac{2-x}{3}\right)}$$ **(2 marks)**

An approximation to the solution of this equation is to be found using the iterative formula

$$x_{n+1} = e^{\left(\frac{2-x_n}{3}\right)},\ x_0 = 1$$

(b) Write down the values of x_1 and x_2, giving your answers to 5 significant figures. **(2 marks)**

(c) Show sufficient working to prove that the x-coordinate of P is 1.274, correct to 4 significant figures. **(3 marks)**

4

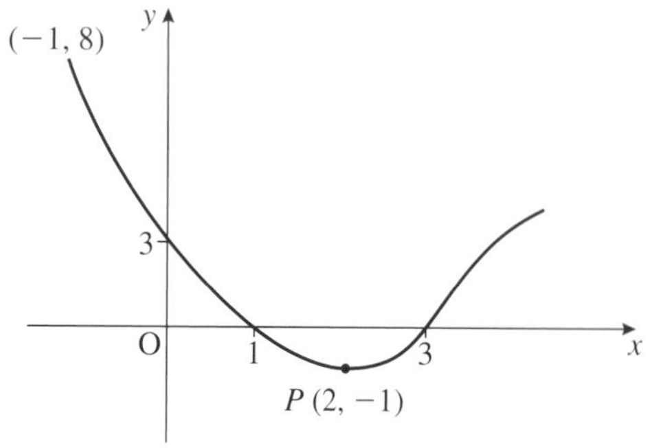

The graph shows part of the curve with the equation $y = f(x)$, $x \in \mathbb{R}$.

The curve passes through the points $(-1, 8)$, $(1, 0)$, $(3, 0)$ and $(0, 3)$. The point $P(2, -1)$ is the only stationary point on the curve.

(a) On separate diagrams, sketch the curve with equation:

(i) $y = 2f(\frac{1}{3}x)$ **(3 marks)**

(ii) $y = f(|x|) + 1$ **(3 marks)**

(iii) $y = |f(x + 1)|$. **(3 marks)**

In each case show the coordinates of points of intersection with the axes.

(b) Find ff(2). **(1 mark)**

5 The function f is defined by

$$f : x \to 3 + \frac{3x + 1}{x^2 + 4x + 3} - \frac{4}{x + 3} \quad \{x \in \mathbb{R}, x > -1\}$$

(a) Show that $f(x) = \dfrac{3x + 2}{x + 1}$, $x > -1$. **(6 marks)**

The function g is defined by

$$g : x \to x^2 - x \quad \{x \in \mathbb{R}\}$$

(b) Find the range of g. **(3 marks)**

(c) Solve $fg(x) = 2$. **(4 marks)**

6 **(a)** The curve C, with equation $y = \sqrt{x} \ln x$, $x > 0$, has a stationary point P. Find, in exact form, the coordinates of P. **(6 marks)**

(b) Given that $y = \tan x + 2 \ln \sec x$, $0 < x < \dfrac{\pi}{2}$, show that

$$\frac{dy}{dx} = (1 + \tan x)^2.$$ **(5 marks)**

7 **(a)** Given that $\cos x = -\frac{1}{3}$, use an appropriate double angle formula to find the exact value of $\sec 2x$. **(4 marks)**

(b) **(i)** Using $\cos (A - B) \equiv \cos A \cos B + \sin A \sin B$, simplify $\cos (x + y) \cos x + \sin (x + y) \sin x$. **(1 mark)**

(ii) Hence, or otherwise, prove that

$$\cos\left(x + \frac{\pi}{3}\right) \operatorname{cosec} x + \sin\left(x + \frac{\pi}{3}\right) \sec x \equiv \operatorname{cosec} 2x,$$

$$\left(x \neq \frac{n\pi}{2}, n \in \mathbb{Z}\right).$$ **(5 marks)**

8 **(a)** Given that $x = \frac{1}{2}\sin y$, $-\frac{\pi}{2} \leqslant y \leqslant \frac{\pi}{2}$, express $\frac{\mathrm{d}y}{\mathrm{d}x}$ in terms of x. **(4 marks)**

The function f is defined by

$$\mathrm{f}: x \rightarrow \tfrac{1}{2}\sin x \quad \left\{x \in \mathbb{R},\ -\frac{\pi}{2} \leqslant x \leqslant \frac{\pi}{2}\right\}$$

(b) Find f^{-1}, and write down the domain of f^{-1}. **(3 marks)**

The curve C has equation $y = \mathrm{f}^{-1}(x)$.

(c) Sketch the graph of C. **(2 marks)**

(d) Explain how your answer in part **(a)** relates to C. **(1 mark)**

Answers to revision questions

Revision exercise 1

1 **(a)** $\frac{5}{2}$ **(b)** Doesn't simplify **(c)** $\dfrac{x+3}{x+4}$

2 **(a)** $\dfrac{3}{xy}$ **(b)** $\dfrac{6}{a}$ **(c)** 1 **3** **(a)** $\dfrac{a}{b}$ **(b)** $2x$ **(c)** $\dfrac{x+2}{x+3}$

4 **(a)** $x^2 + 4x + 5 + \dfrac{5}{x-2}$ **(b)** $4x - 6 + \dfrac{21x - 24}{x^2 + x - 3}$

5 $3x^2 + x + 1 + \dfrac{6}{x-2}$ **6** Proof

7 **(a)** $x^2 + 2x - 3 = (x+3)(x-1)$ **(b)** $x = 2$

$$\begin{aligned} f(x) &= \frac{x(x^2 + 2x - 3) + 3(x+3) - 12}{(x+3)(x-1)} \\ &= \frac{x^3 + 2x^2 - 3}{(x+3)(x-1)} \\ &= \frac{(x-1)(x^2 + 3x + 3)}{(x-1)(x+3)} \\ &= \frac{(x^2 + 3x + 3)}{(x+3)} \end{aligned}$$

Revision exercise 2

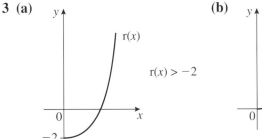

1 **(a)** 8 **(b)** 2.5 **2** **(a)** 1.75 **(b)** 8

3 **(a)**

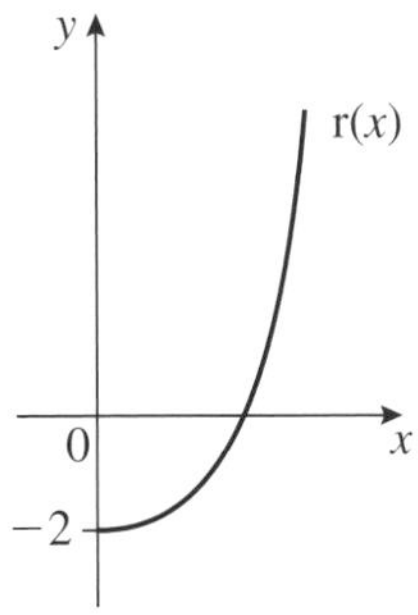

r(x) > −2

(b)

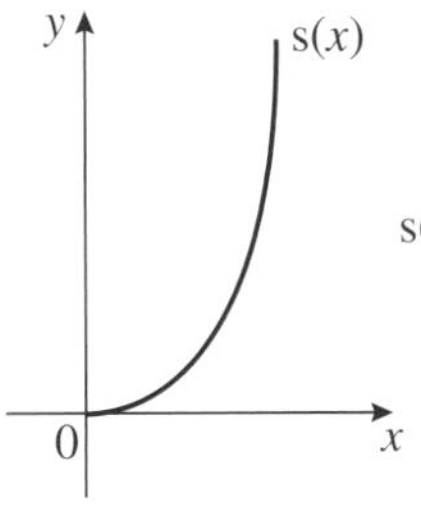

s(x) ⩾ 0

3 **(c)**

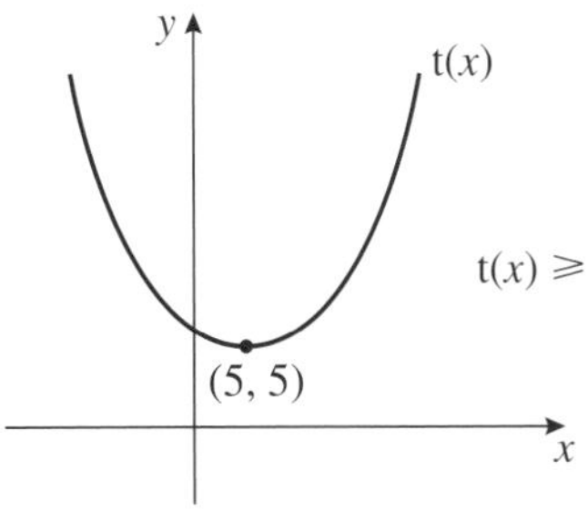

t(x) ⩾ 5

4 f(x) is not a function because it is NOT defined at $x = 0$

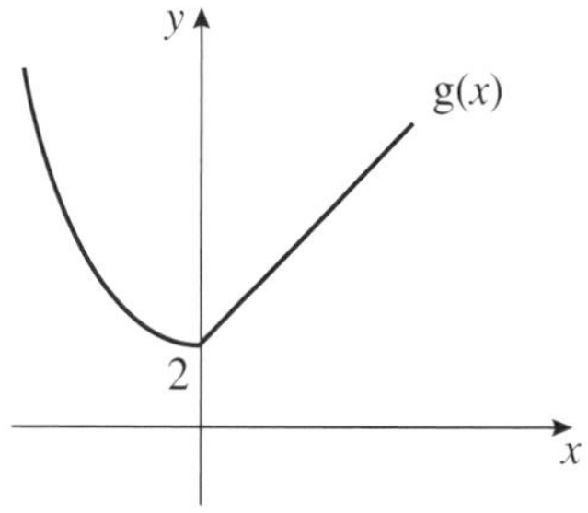

(a) 6 **(b)** 11 **(c)** −6, 12

5 (a) $2x^2 + 6$ **(b)** $4x^2 + 16x + 17$ **(c)** $4x + 12$ **(d)** 3, −1

6 (a) $f^{-1}(x) = \dfrac{x-5}{2}$ **(b)** $g^{-1}(x) = \sqrt[3]{\dfrac{x+5}{2}}$

6 (c) $h^{-1}(x) = \dfrac{1}{x} - 2, \quad x \neq 0$

7 (a) $p(x) \geqslant -3$ **(b)** $p^{-1}(x) = \sqrt{\dfrac{x+3}{2}}, x \geqslant -3$

7 (c)

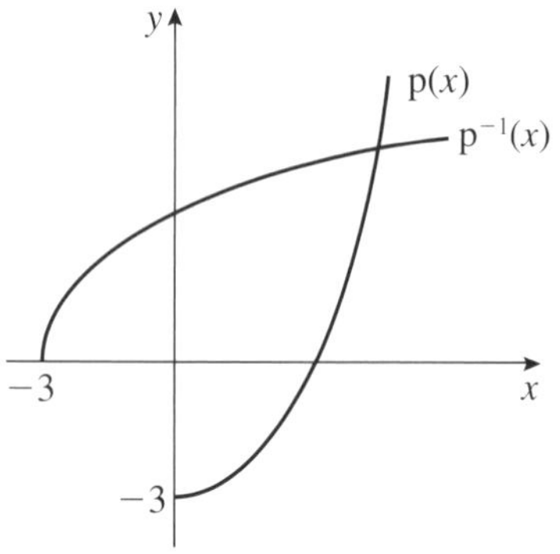

$p^{-1}(x)$ is a reflection of $p(x)$ in the line $y = x$

7 (d) $a = 1.5$

8 (a) $g^{-1}(x) = \dfrac{x+5}{x-2}$ $\{x \in \mathbb{R}, x > 2\}$

8 (b) $g^{-1}(x) > 1$ **(c)** $x > 2$

9 (a) $x = 6, x = -2$ **(b)** Range: $g(x) \geqslant 7$ **(c)** $gf(-1) = 11$

Revision exercise 3

1 (a)

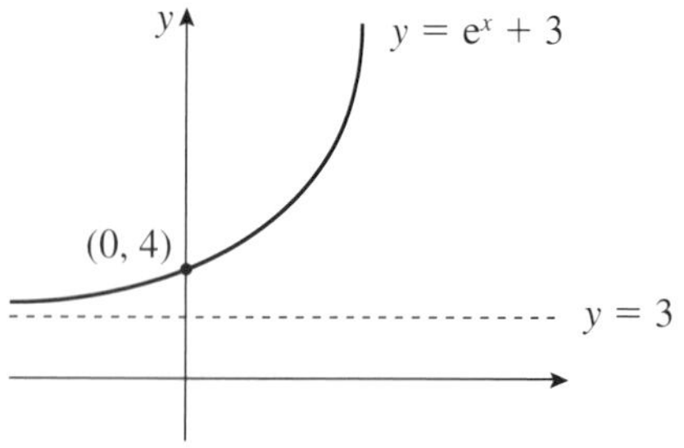

1 (b)

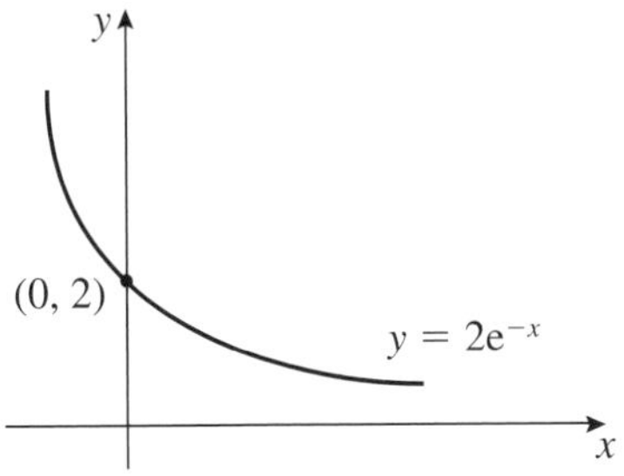

1 (c)

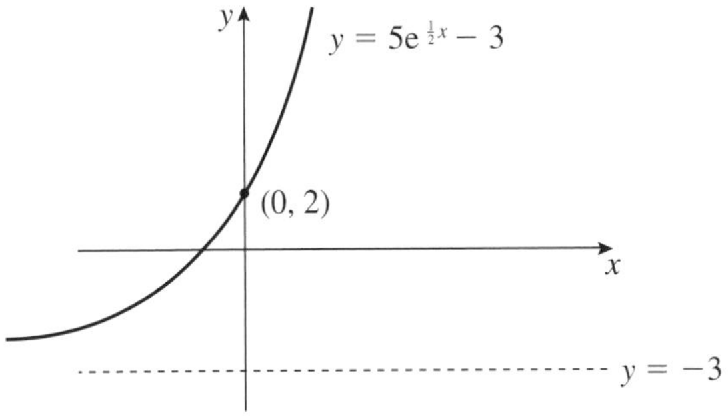

2 (a) $A = (0, -2)$ $B = (\frac{1}{2}\ln(\frac{5}{3}), 0)$ **(b)** $f(x) > -5$

2 (c) $f^{-1}(x) = \frac{1}{2}\ln\left(\frac{x+5}{3}\right), x > -5$

3 (a) $e^{\frac{3}{2}}$ **(b)** $\ln 2$ **(c)** $\dfrac{e^8 - 3}{2}$ **(d)** $\ln 2$

4 (a) $A = (-5 + e^{-3}, 0)$ $B = (0, 3 + \ln 5)$ **(b)** $f^{-1} : (x) \mapsto e^{x-3} - 5$

4 (c)

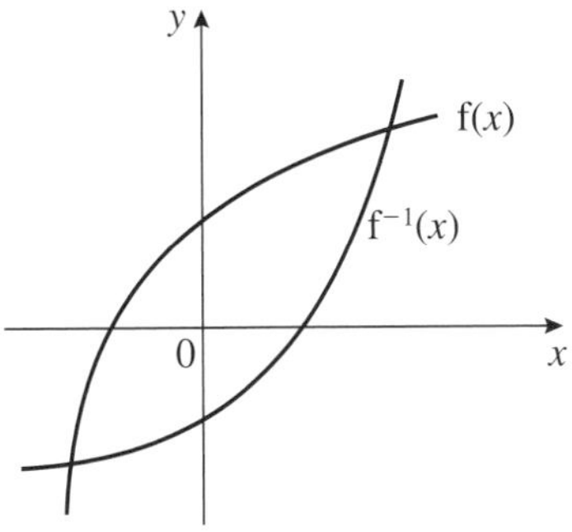

(d) 2 roots

5 (a) $g(x) > 0$ **(b)** 1 **(c)**

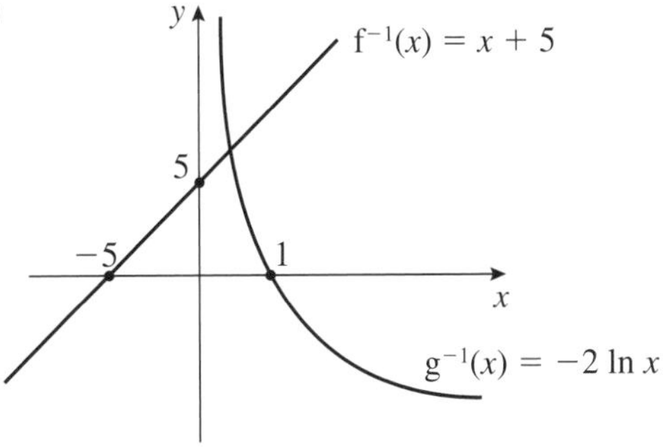

6 (a) $f(x) > -k$ **(b)** 0 **(c)** 5

7 (a) 80 **(b)** 47 **(c)** 2022 **(d)** 200

8 (a) $h(x) \leqslant 70$ **(b)** $A = \left(\frac{3 \ln 2}{5}, 0\right)$ **(c)** $h^{-1}(x) = \frac{1}{5} \ln \left(\frac{80 - x}{10}\right), x \leqslant 70$

8 (d)

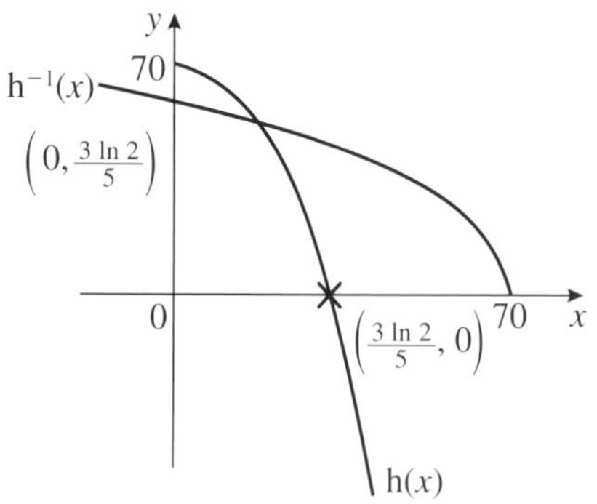

$h^{-1}(x)$ is a reflection of $h(x)$ in the line $y = x$

9 (a) $T = 80$ **(b)** $e^{-0.1t} > 0$ **(c)**

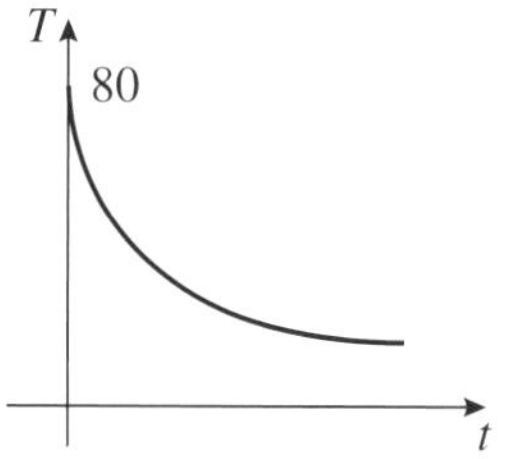

(b) $t = 4.1$

Revision exercise 4

1 $(-4, -3), (2, 3), (5, 6)$

2 (a) $f(3) = -7, f(4) = 18$

3 (a) $a = 3, b = 1$ **(b)** 1.307

4 (a) 2 roots

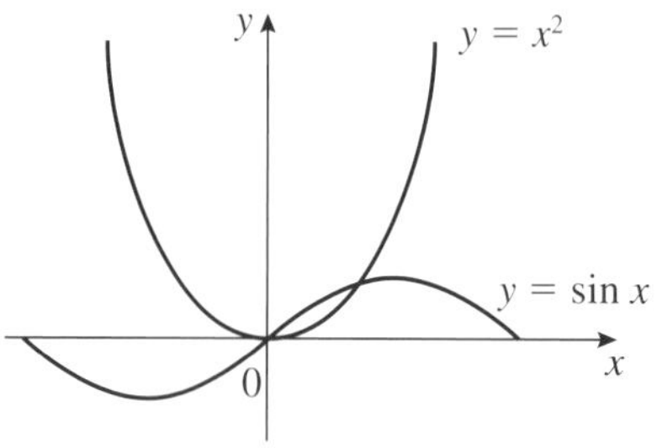

4 (b) $\sin(0.8) - 0.8^2 = 0.077, \sin(0.9) - 0.9^2 = -0.027$ **(c)** 0.877

5 (a) $f(4.5) = -0.108, f(4.6) = 3.148$ **(c)** 4.503

6 (a)

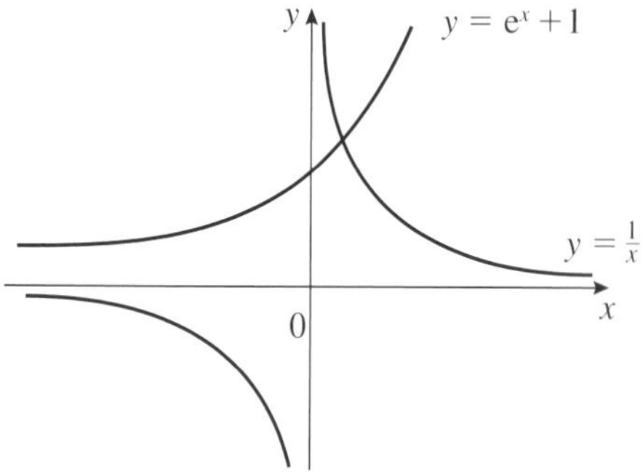

6 (b) $e^{0.4} + 1 - \dfrac{1}{0.4} = -0.008$, $e^{0.41} + 1 - \dfrac{1}{0.41} = 0.068$ **(c)** 0.401

7 (a) 0.473 **(b)** $y^3 + 4y - 2 = 0$ **(c)** -1.1

8 (a) $f(1.2) = -0.056$, $f(1.21) = 0.091$ **(b)** $a = 0.5$, $b = 2$

8 (c) 1.204

9 (a) $f\left(\dfrac{\pi}{2}\right) = \dfrac{\pi}{4} - 1$, $f(\pi) = \dfrac{\pi}{2} - 1$

9 (b) Converges to a root outside $\dfrac{\pi}{2} < x < \pi$.

10 (a)

x	-3	-2	-1	0	1	2	3
$f(x)$	-28.6	-9.74	-3	-3	-3	3.74	22.6

(b) 1 **(c)** 1.61

11 (a) $f(2) = 8 - 4 - 5 = -1$, $f(3) = 27 - 6 - 5 = 16$

11 (b) $x_1 = 2.121$, $x_2 = 2.087$, $x_3 = 2.097$, $x_4 = 2.094$

11 (c) Choosing suitable interval, eg [2.09455, 2.09465]
$f(2.09455) = -0.00001\ldots$
$f(2.09465) = +0.00110\ldots$

12 (a) $4.5e^{0.5x} - 3x^2$ **(b)** $f'(2) = 0.232$, $f'(2.1) = -0.371$

12 (c) 2.04

13 (a) $7 - 2x - 3e^{-x}$ **(c)** -1.126 **(d)** 0.100

14 (b) 1.58, 1.68, 1.70 **(c)** $f(1.695) = -0.037$, $f(1.705) = 0.084$

14 (d) Any x_1 with $-1 \leqslant x_1 \leqslant -\frac{1}{4}$, as we either have to divide by 0 or take the square root of a negative number

Revision exercise 5

1

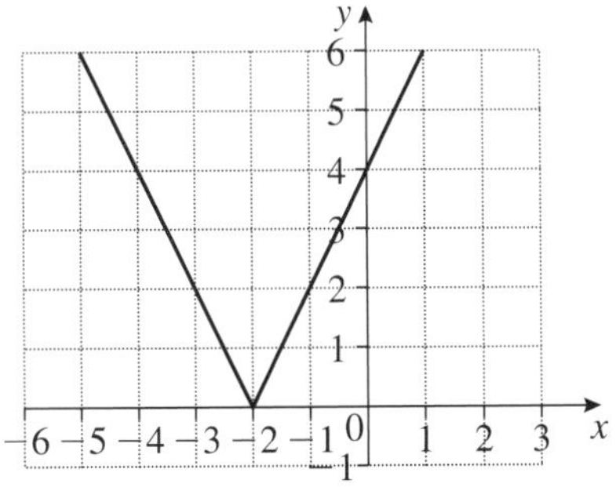

$(-2, 0)$, $(0, 4)$

2

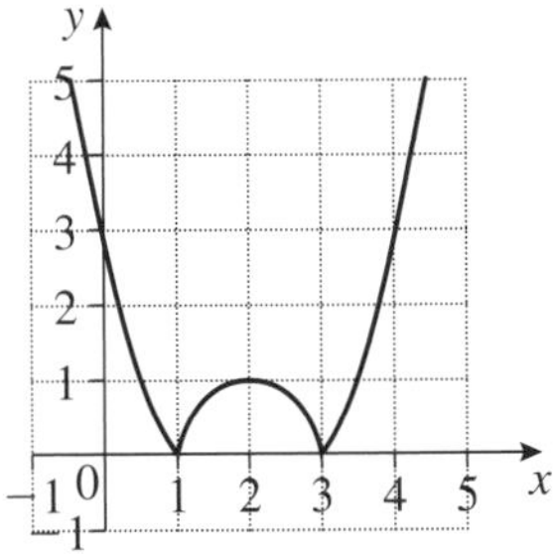

$(1, 0)$, $(3, 0)$, $(0, 3)$

3

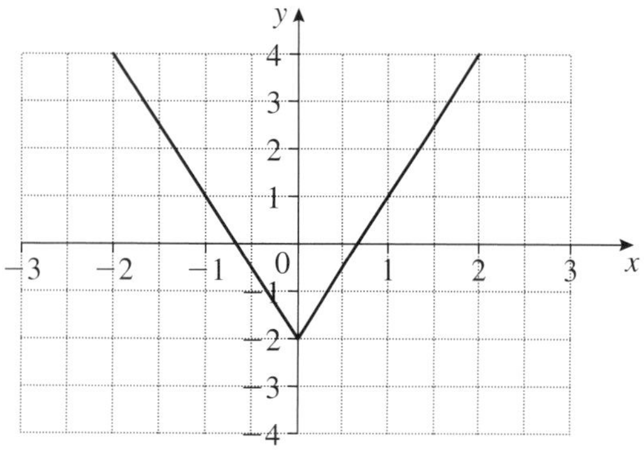

$(-\frac{2}{3}, 0), (\frac{2}{3}, 0), (0, -2)$

4 (a)

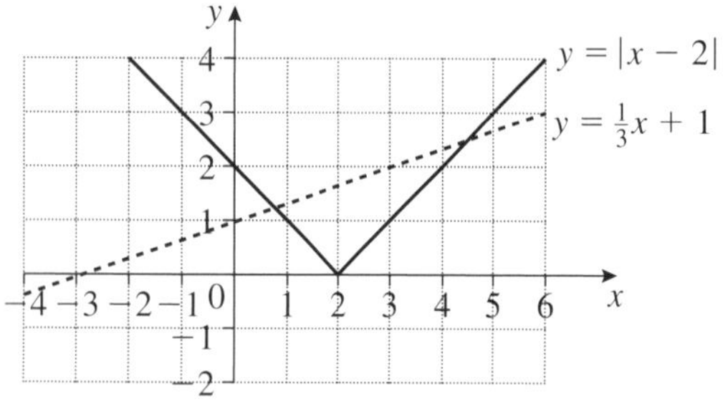

(b) $x = \frac{3}{4}, x = \frac{9}{2}$

5 (a)

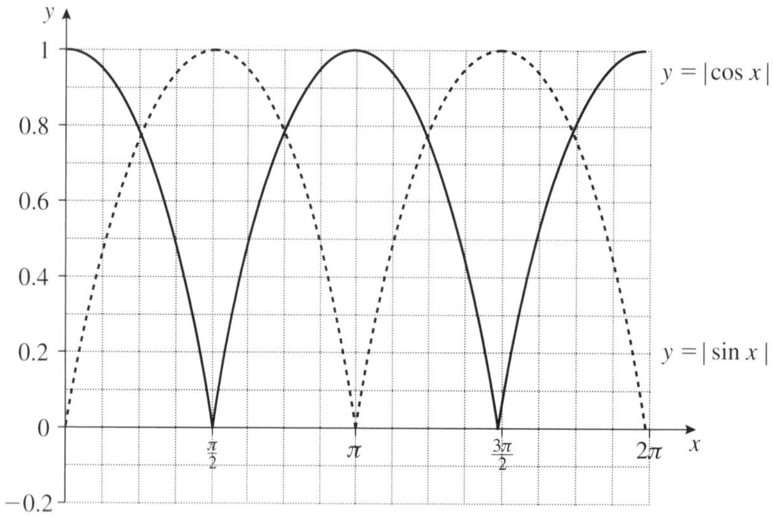

5 (b) $\left(\frac{\pi}{4}, \frac{1}{\sqrt{2}}\right), \left(\frac{3\pi}{4}, \frac{1}{\sqrt{2}}\right), \left(\frac{5\pi}{4}, \frac{1}{\sqrt{2}}\right), \left(\frac{7\pi}{4}, \frac{1}{\sqrt{2}}\right)$

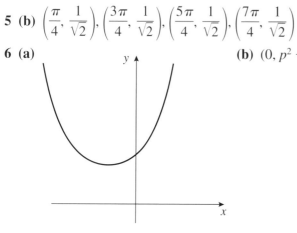

6 (a)

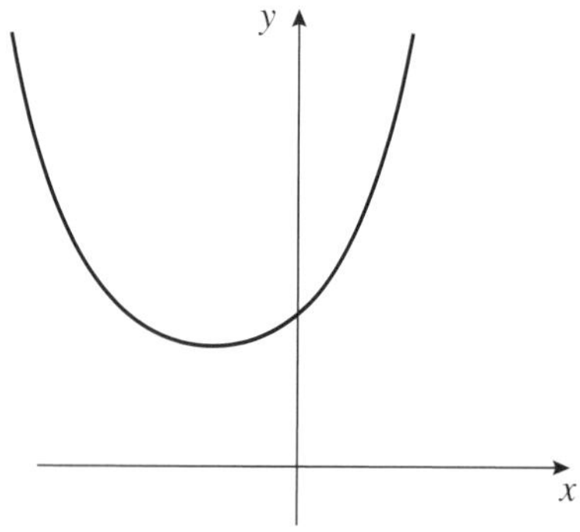

(b) $(0, p^2 + q)$ **(c)** $(-p, q)$

7 **(a)**

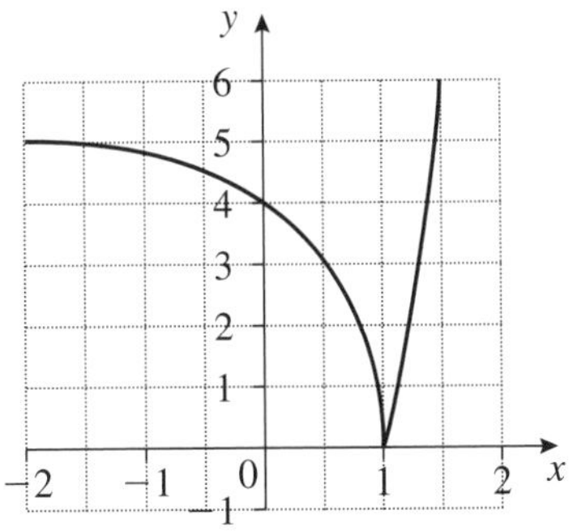

(1, 0), (0, 4) **(b)** 0.68, 1.21

8 **(a)**

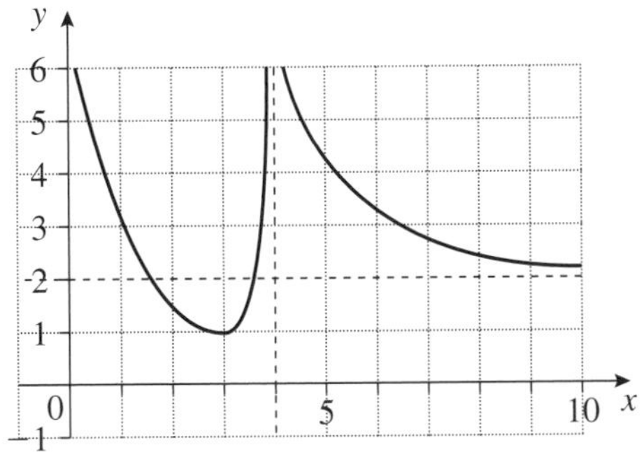

(b) (3, 1) **(c)** $x = 0, x = \ 4, y = 2$

9 **(a)**

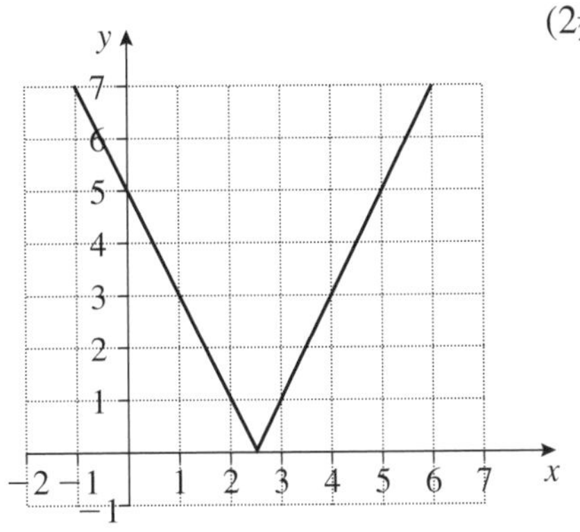

$(2\frac{1}{2}, 0), (0, 5)$

(b) $x = \frac{5}{3}, x = 3$ **(c)** $g(x) \geqslant -9$ **(d)** 15

10 **(a)**

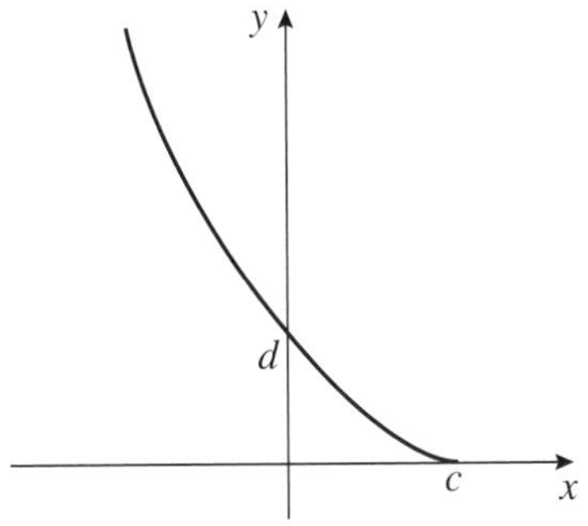

$(c, 0), (0, d)$

10 (b)

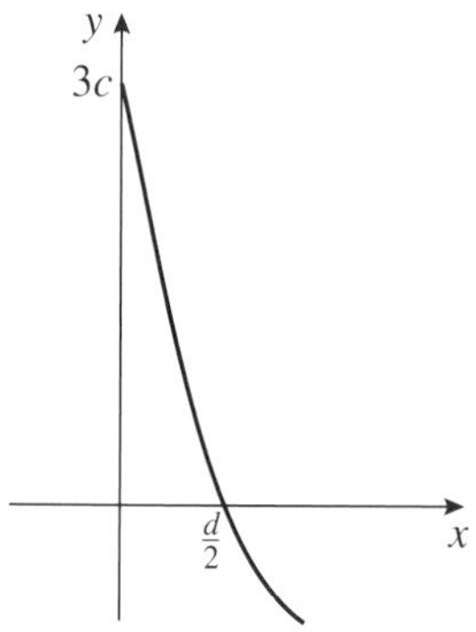

$\left(\frac{d}{2}, 0\right)$, (0. 3$c$)

(c) (i) $c = 2$ (ii) $-1 < \mathrm{f}(x) \leqslant 2$ **(d)** 1.585 **(e)** $\frac{3}{x} - 1$

Revision exercise 6

1 (a) $\cot^2 \theta$ **(b)** $\cos 2\theta$ **(c)** $\cot^2 \theta$ **(d)** $\sin \theta$ **(e)** $\sec^4 \theta$ **(f)** 1

2 (a) -1 **(b)** $\sqrt{3}$ **(c)** $\sqrt{2}$ **(d)** -2

3 (a) 150°, 330° **(b)** 30°, 150°, 210°, 330° **(c)** 63.4°, 243.4°

3 (d) 0, 45°, 180°, 225°, 360° **(e)** 52.5°, 142.5°, 232.5°, 322.5°

3 (f) 35.3°, 144.7°, 215.3°, 324.7° **(g)** 0°, 90°, 180°, 270°, 360°

4 (a)

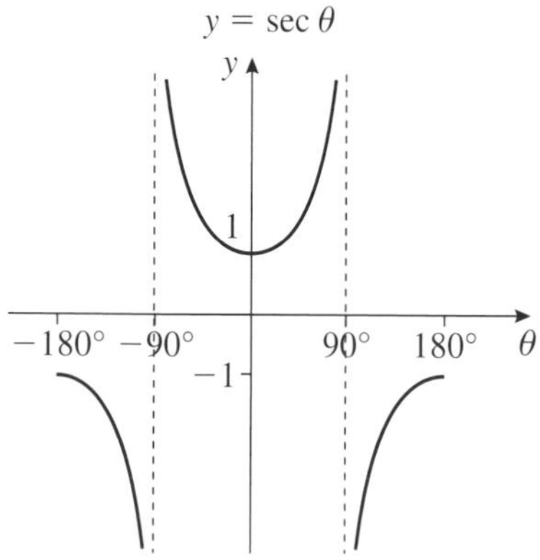

4 (b)

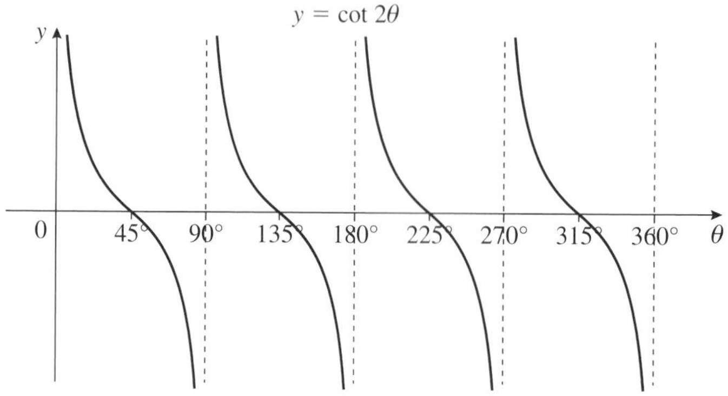

4 (c)

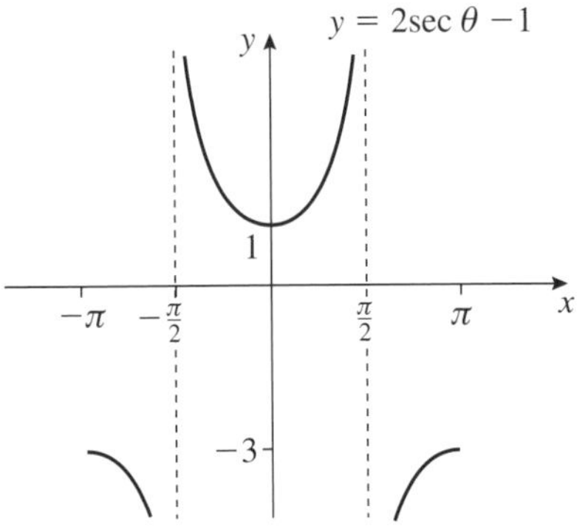

4 (d)

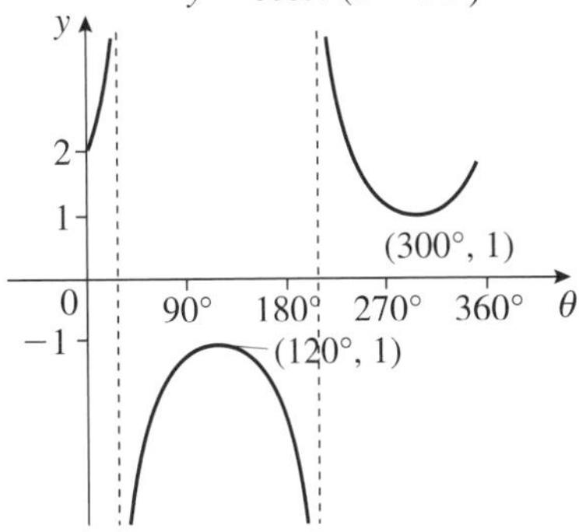

5 (a) LHS $= (1 - \cos^2\theta) + (\sec^2\theta - 1) \equiv \sec^2\theta - \cos^2\theta = \text{RHS}$

5 (b) LHS $= \dfrac{\sin\theta}{\cos\theta} + \dfrac{\cos\theta}{\sin\theta} \equiv \dfrac{\sin^2\theta + \cos^2\theta}{\sin\theta\cos\theta} \equiv \dfrac{1}{\sin\theta\cos\theta}$

$$\equiv \operatorname{cosec}\theta\sec\theta = \text{RHS}$$

5 (c) LHS $= (1 + \tan^2\theta)^2 - \tan\theta \equiv 1 + \tan^2\theta + \tan^4\theta$

$$\equiv \sec^2\theta + \tan^4\theta = \text{RHS}$$

5 (d) LHS $= \dfrac{\cos\theta(\operatorname{cosec}\theta + 1) + \cos\theta(\operatorname{cosec}\theta - 1)}{(\operatorname{cosec}\theta - 1)(\operatorname{cosec}\theta + 1)}$

$$\equiv \frac{\cot\theta + \cos\theta + \cot\theta - \cos\theta}{\operatorname{cosec}^2\theta - 1}$$

$$\equiv \frac{2\cot\theta}{\cot^2\theta} \equiv \frac{2}{\cot\theta} \equiv 2\tan\theta = \text{RHS}$$

6 (a) $\dfrac{1}{p}$ **(b)** $\dfrac{1}{q^2} - 1$ **(c)** $\dfrac{p}{q}$

7 (a) 0 **(b)** $\dfrac{\sqrt{3}}{2}$ **(c)** $\dfrac{1}{2}$

8 $\dfrac{13\pi}{12}$

9 **(a)** $\cos A = \dfrac{2^2 + 5^2 - 6^2}{2 \times 5 \times 2} = -\dfrac{7}{20}$, so $\sec A = -\dfrac{20}{7}$

(b) $\dfrac{\sin B}{2} = \dfrac{\sin A}{6}$, so $\sin B = \dfrac{\sin A}{3}$

As $\cos A = -\dfrac{7}{20}$, $\sin A = \dfrac{\sqrt{351}}{20}$, so $\sin B = \dfrac{\sqrt{351}}{60}$ and $\operatorname{cosec} B = \dfrac{60}{\sqrt{351}}$

10 **(a)** $\dfrac{\pi}{12}, \dfrac{5\pi}{12}, \dfrac{3\pi}{4}$ **(b)** 77.9°, 282.0°

11 **(a)** $-\dfrac{\pi}{2}, \dfrac{\pi}{6}, \dfrac{5\pi}{6}$ **(b)** 35.8°, 125.8°, 215.8°, 305.8°

12 $\arccos \frac{1}{3}$ is the angle α, $0 \leqslant \alpha \leqslant \pi$, such that $\cos \alpha = \frac{1}{3}$

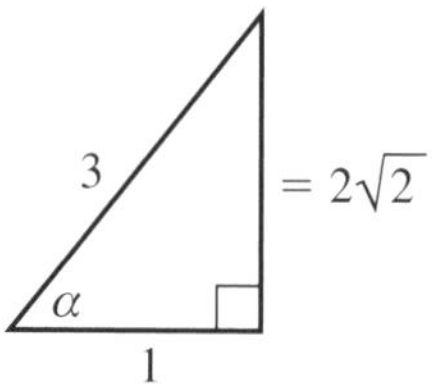

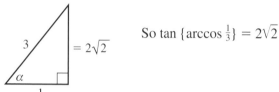

So $\tan \{\arccos \frac{1}{3}\} = 2\sqrt{2}$

13 **(a)** $\frac{4}{3}$ **(b)** $-\frac{5}{4}$ **14** **(a)** $-2\sqrt{2}$ **(b)** $-\dfrac{\sqrt{2}}{4}$ **(c)** $-\dfrac{3\sqrt{2}}{4}$

15 **(a)** $\dfrac{3}{5 - 4 \cot x}$ **16** **(a)** $\dfrac{\pi}{2}$

17 **(a)** $(a - 1)(b - 1)$ **(b)** $\dfrac{\pi}{4}, \dfrac{\pi}{2}, \dfrac{5\pi}{4}$

18 **(a)** 0.841 **(b)**

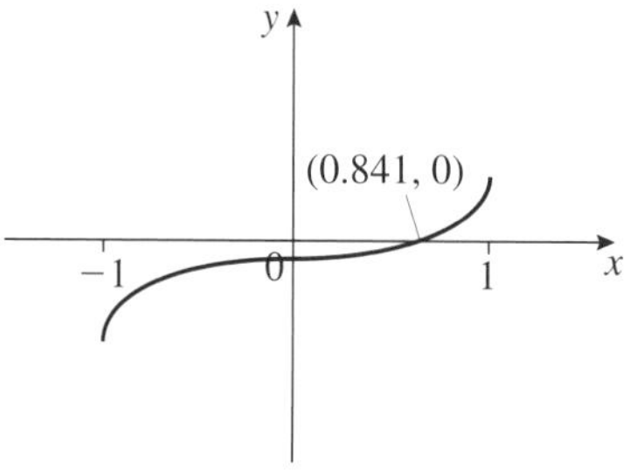

19 **(a)**

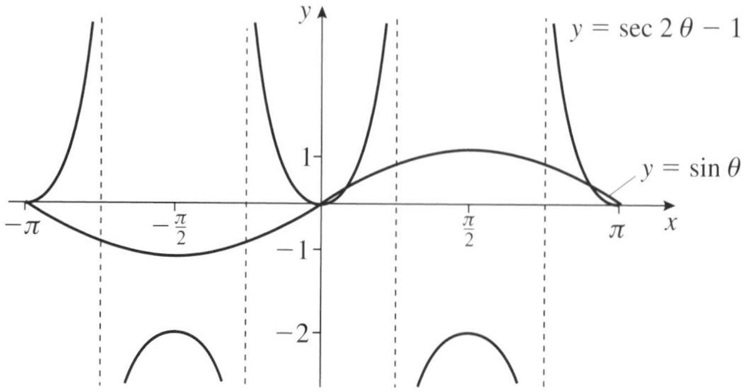

19 **(b)** $-\pi, 0, 0.375, \pi, 2.77$

20 $-150, -30, 90$

Revision exercise 7

1 (a) (d)

2 (a) (i) $\sin 60°$ **(ii)** $\cos 150°$ **(iii)** $\tan 45°$

2 (b) (i) $\frac{\sqrt{3}}{2}$ **(ii)** $-\frac{\sqrt{3}}{2}$ **(iii)** 1

3 (a) $\text{RHS} = \frac{1 + (2\cos^2 2\theta - 1)}{2} \equiv \frac{2\cos^2 2\theta}{2} \equiv \cos^2 2\theta = \text{LHS}$

3 (b) $\text{LHS} = \cos^2\theta + 2\sin\theta\cos\theta + \sin^2\theta - (\cos^2\theta - 2\sin\theta\cos\theta + \sin^2\theta)$

$\equiv (\cos^2\theta - \cos^2\theta) + (\sin^2\theta - \sin^2\theta) + 4\sin\theta\cos\theta$

$\equiv 2\sin 2\theta = \text{RHS}$

4 $\cos\theta = x - 1$, $\cos 2\theta = y$ so $y = 2(x-1)^2 - 1 \Rightarrow y = 2x^2 - 4x + 1$

5 $\frac{k+1}{1-k}$ **6** $\frac{3 - \tan x}{1 - 3\tan x}$ **7 (a)** $\frac{3\pi}{4}$ **(b)** $0, \frac{\pi}{6}, \frac{5\pi}{6}, \pi$ **(c)** $\frac{13\pi}{15}$

8 $-180°, -80.4°, 0°, 80.4°, 180°$

9 (a) $\frac{4}{5}$ **(b)** $\frac{25}{24}$ **(c)** $\frac{7}{9}$ **(d)** $\frac{8\sqrt{2} - 3}{15}$

10 $c = 2\sqrt{3}$, $k = 4$

12 (a) $\pm\frac{\sqrt{3}}{6}$ **(b) (i)** $-\frac{1}{3}$ **(ii)** $-\frac{\sqrt{10}}{3}$ **13 (b) (i)** $-\frac{63}{16}$ **(ii)** $-\frac{33}{65}$

14 (a) $\text{LHS} = \frac{\cos 2\theta}{\sin 2\theta} - \frac{\sin 2\theta}{\cos 2\theta} \equiv \frac{\cos^2 2\theta - \sin^2 2\theta}{\sin 2\theta \cos 2\theta} = \frac{\cos 4\theta}{\frac{1}{2}\sin 4\theta}$

$\equiv \frac{2\cos 4\theta}{\sin 4\theta} = \text{RHS}$

14 (b) $\text{LHS} = 1 + 2\cos^2\theta - 4(1 - \cos^2\theta)$

$= 6\cos^2\theta - 3$

$= 3(2\cos^2\theta - 1)$

$= 3\cos 2\theta = \text{RHS}$

14 (c) $\text{LHS} = \frac{\sin A}{\cos A} + \frac{\cos B}{\sin B}$

$\equiv \frac{\sin A \sin B + \cos A \cos B}{\cos A \sin B}$

$\equiv \sin(A + B) \times \frac{1}{\cos A} \times \frac{1}{\sin B}$

$\equiv \sin(A + B)\sec A \operatorname{cosec} B = \text{RHS}$

15 (a) $111.5°, 248.5°$ **(b)** $126.2°, 306.2°$ **(c)** $0°$, 26,6°, $206.6°, 180°$

16 (b)

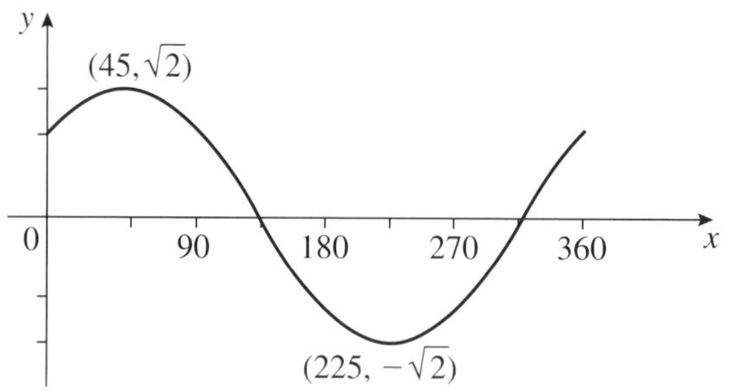

17 (a) $\sqrt{41}\cos(x - 38.7°)$ **(b)** $110.5°, 326.9°$

18 (a) $-56.3°, 123.7°$ **(b)** $-31.7°, 99.1°$

18 (c) $-146.0°, -34.0°, 63.3°, 116.7°$

19 (b) $$\text{LHS} = \frac{2\cos 2x \cos x}{-2\sin 2x \sin(-x)}$$
$$\equiv \frac{2\cos 2x \cos x}{2\sin 2x \sin x}$$
$$\equiv \frac{\cos 2x}{\sin 2x} \cdot \frac{\cos x}{\sin x}$$
$$\equiv \cot 2x \cot x = \text{RHS}$$

19 (c) $x = 30°, 45°, 90°, 135°, 150°$
[$\cos 5x + \cos x = 0 \Rightarrow 2\cos 3x \cos 2x = 0$, so $\cos 3x = 0$ or $\cos 2x = 0$]

20 (b) $$\text{LHS} = \frac{2\sin 30° \cos(-15°)}{2\cos 30° \cos(-15)°} = \tan 30° = \frac{\sqrt{3}}{3}$$

Revision exercise 8

1 (a) $-6e^{-0.1t}$ **(b)** 38

2 $-\frac{6}{5}$

3 $(-2, -\frac{1}{4}), (2, \frac{1}{4})$

4 $-8, 2$

5 (a) $(y + 4)e^y$ **(b)** $\frac{1}{4}$

6 (a) $x + 2y = 14$ **(b)** When $y = 0$, $x + 2y = 14 \Rightarrow x = 14$.

7 (a) $1 + \dfrac{e^x}{5}$ (b) $y = \frac{6}{5}x + \frac{1}{5}$

8 (a) $\frac{1}{3}$ **(b)** $y = 3\ln\frac{1}{3} + \dfrac{1}{\frac{1}{3}} = 3 - 3\ln 3$

9 $f'(x) = -\dfrac{(x-1)^2}{x^2+1} \leqslant 0$, so f is decreasing

10 $\dfrac{16 + \sqrt{3}}{2}$

13 $\frac{dy}{dx} = -2e^{-2x}\cos 3x - 3e^{-2x}\sin 3x$

$\frac{d^2y}{dx^2} = 13e^{-2x}\sin(3x - 0.395)$

14 $y = x - \frac{\pi}{4} - \frac{1}{2}\ln 2, \left(\frac{\pi}{4} + \frac{1}{2}\ln 2, 0\right)$

Examination style paper

1 (a) 25 **(b)** 189.9 or 190 **(c)** 0.133

2 193.0°, 347.0°

3 (a) Meet where $2 - x = 3\ln x \Rightarrow \ln x = \left(\frac{2-x}{3}\right) \Rightarrow x = e^{\left(\frac{2-x}{3}\right)}$

3 (b) $x_1 = 1.3956$, $x_2 = 1.2232$

3 (c) Setting up relevant function, e.g. $f(x) = 2 - x - 3\ln x$
Evaluating $f(x)$ for two values in [1.2735, 1.2745]
Showing change of sign and conclusion

4 (a) (i)

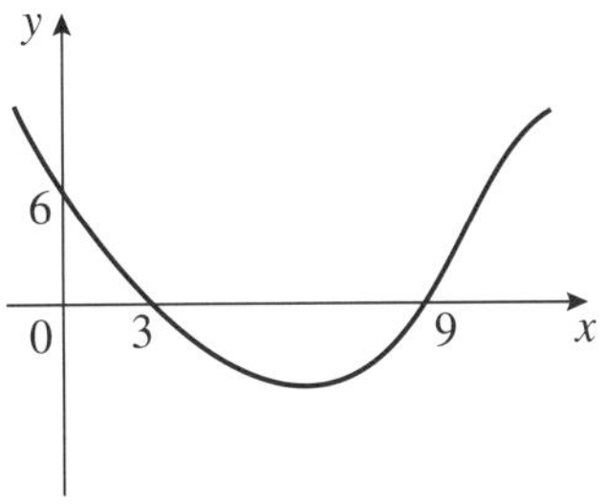

(ii)

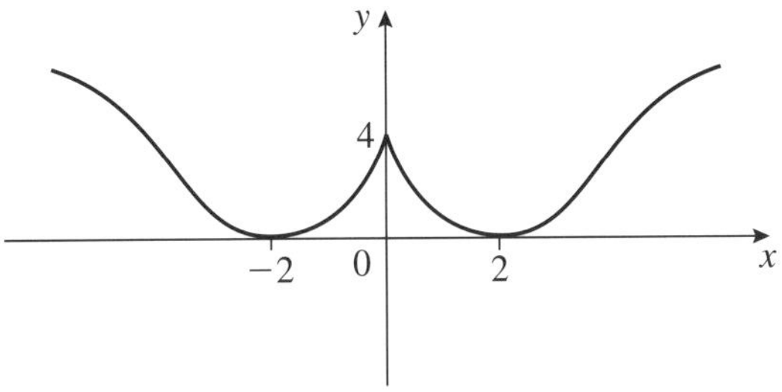

(ii)

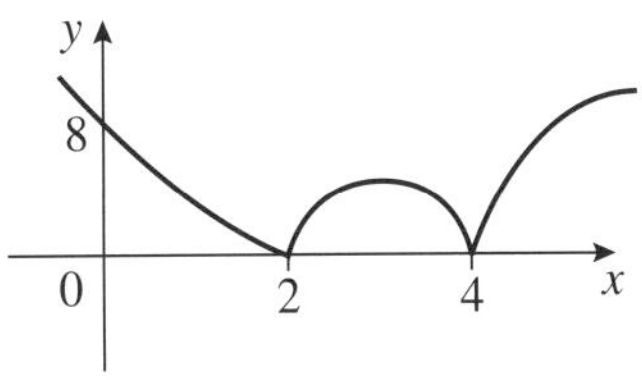

4 (b) 8

5 (a) $f(x) = \frac{3(x^2 + 4x + 3) + (3x + 1) - 4(x + 1)}{(x + 3)(x + 1)}$

$= \frac{3x^2 + 11x + 6}{(x + 3)(x + 1)} = \frac{(3x + 2)(x + 3)}{(x + 3)(x + 1)} = \frac{(3x + 2)}{(x + 1)}$

5 (b) $g(x) \geqslant -\frac{1}{4}$ **(c)** $x = 0, 1$

6 (a) $x = e^{-2}, y = -2e^{-1}$

6 (b) $\dfrac{dy}{dx} = \sec^2 x + \dfrac{2\sec x \tan x}{\sec x} = 1 + \tan^2 x + 2\tan x = (1 + \tan x)^2$

7 (a) $-\frac{9}{7}$

7 (b) (i) $\cos y$

(ii) $\text{LHS} = \dfrac{\cos\left(x + \dfrac{\pi}{3}\right)}{\sin x} + \dfrac{\sin\left(x + \dfrac{\pi}{3}\right)}{\cos x}$

$$= \frac{\cos\left(x + \dfrac{\pi}{3}\right)\cos x + \sin\left(x + \dfrac{\pi}{3}\right)\sin x}{\sin x \cos x}$$

$$= \frac{\cos\left(\dfrac{\pi}{3}\right)}{\sin x \cos x} = \frac{1}{2\sin x \cos x} = \operatorname{cosec} 2x$$

8 (a) $\dfrac{2}{\sqrt{1 - 4x^2}}$

(b) $f^{-1} : x \to \arcsin 2x \ (\sin^{-1} 2x)$

Domain: $-\frac{1}{2} \leqslant x \leqslant \frac{1}{2}$

(c)

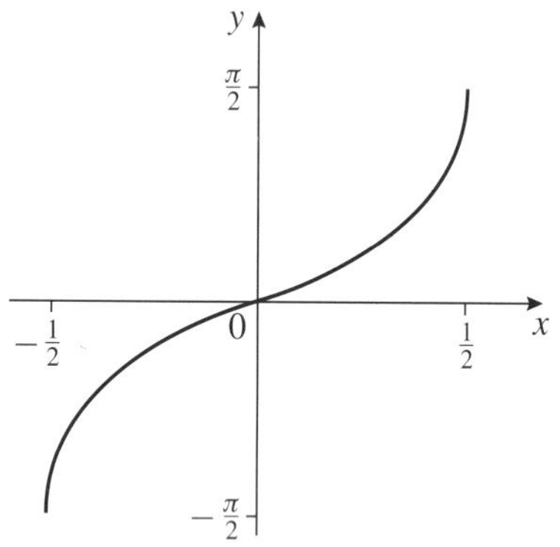

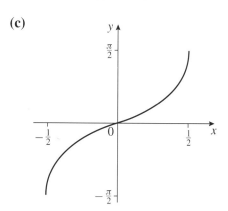

(d) If $x = \frac{1}{2}\sin y$, $-\dfrac{\pi}{2} \leqslant y \leqslant \dfrac{\pi}{2}$, then $y = \arcsin 2x$, $-\frac{1}{2} \leqslant x \leqslant \frac{1}{2}$ and so they represent the same curve.